Lecture Notes in Computer Science

Edited by G. Goos and J. Hartmanis

D0992923

370

Christoph Meinel

Modified Branching Programs and Their Computational Power

Springer-Verlag

Berlin Heidelberg New York London Paris Tokyo Hong Kong

Author

Christoph Meinel
Sektion Mathematik, Humboldt-Universität Berlin
PF 1297, Unter den Linden, DDR-1056 Berlin, GDR

CR Subject Classification (1987): F.1.1–3, F.2.2, G.2.2

ISBN 3-540-51340-X Springer-Verlag Berlin Heidelberg New York
ISBN 0-387-51340-X Springer-Verlag New York Berlin Heidelberg

PREFACE

This monograph is a revised version of my habilitation thesis submitted to the presidential council of the Academy of Science of GDR. It summarizes the results of my research on branching program based complexity theory over the last years.

Many people contributed to this work. First of all I want to thank Lothar Budach. His confidence, criticism, and encouragement have been very important for me.

Special thanks are due to my colleagues and friends Stephan Waack and Matthias Krause for many fruitful discussions on the subject of this paper. I benefitted highly from discussions with Ingo Wegener and from his book "The Complexity of Boolean Functions".

Thanks go also to Prof. G.Hotz and Prof. G.Asser and the known and unknown referees of the journals "Theoretical Computer Science", "Information and Computation", "Journal of Information Processing and Cybernetics (EIK)", "Fundamenta Informaticae" and to the members of the program committees of FCT, MFCS and STACS for the appreciation of my work as well as for hints and comments.

Ivanka deserves special thanks for her love and patience.

Berlin, April 1989   Ch. M.

CONTENTS

εχ μερου   γαρ   γινωσχομεν

(*for we know in part*)

N.T. , 1 Cor. 13.9

INTRODUCTION

One of the fundamental issues of complexity theory is to estimate the relative efficiency of different models of computation. A general program in doing this has been to take abstract models of computation, such as Turing machines, Random Access Machines, Boolean circuits or branching programs, and examine their behavior under certain resource constraints. This leads to the definition of complexity classes which formalize certain computational powers. By examining the meaning of and the relationships between such classes, one seeks to understand the relative strengths of their underlying computational paradigms.

In recent years the concepts of Boolean circuit complexity and other circuit based *nonuniform* complexities have been (re-)discovered to be useful in complexity theoretic research. They are based on computational models which are purely combinatorial objects. Apart from the strong practical interest in investigations of such circuit models, nonuniform complexity classes appear to be more amenable to combinatorial analysis. Results obtained by Furst, Saxe and Sipser [FSS84], Razborov [Ra85], [Ra86], Andreev [An85], Yao [Ya85], Hastad [Ha86], Barrington [Ba86] and many others have advanced our knowledge of nonuniform complexity classes as well as of complexity classes in general.

One of the most important nonuniform models of computation

are *branching programs* which generalize the concept of decision trees. The settings of n input variables determine a flow of control through a branching program, as each node activates one of two successors depending on the value of a tested input bit. Originally invented for the analysis of switching problems, branching programs have come to be analyzed as an abstract model of computation.

In this thesis we examine modified branching programs and their computational power. The most natural measure for the branching program complexity of a (Boolean) function is the *size* of a minimal branching program which computes this function. The well-known relations between branching program size and the space complexity of nonuniform Turing machines [Co66], [Bu82], [PŽ83] were the starting-point of our investigations.

First, we introduce and examine *nondeterministic branching programs*. It is found that the computational power of nondeterministic branching programs is mainly determined by their ability to record nondeterministic choices. While the size of 1-time-only-nondeterministic branching programs is related to nondeterministic (nonuniform) Turing machine space, the size of unrestricted nondeterministic branching programs is related to nondeterministic (nonuniform) Turing machine time [Me86,2]. Furthermore, nondeterministic and 1-time-only-nondeterministic branching programs respond differently to bounding their width. While nondeterministic bounded width branching programs turn out to be as powerful as nondeterministic branching programs without any width constraints, 1-time-only-nondeterminism does not even increase the computational power of (ordinary) bounded width branching programs.

We then generalize the concept of 1-time-only-nondeterministic branching programs by introducing Ω-*branching programs*. Ω-branching programs are branching programs some of whose nodes are capable of evaluating Boolean functions from a set $\Omega \subseteq \mathbf{B}_2$ of 2-argument Boolean functions. We prove that complexity classes defined by Ω-branching programs fit closely into the framework of already studied classes defined by Turing machines or Boolean circuits. We completely classify Ω-branching programs into *ordinary, disjunctive, conjunctive, parity and alternating branching programs*, and relate the size of these branching programs to the space of nonuniform deterministic, nondeterministic, co-nondeterministic, "parity" and alternating Turing machines, respectively. Hence, four of the five types of Ω-branching programs correspond to well-known types of Turing machines whereas the fifth type has not been identified up to now in the context of space-bounded Turing machines.

The study of the influence of certain resource constraints on Ω-branching programs yields interesting results also for the corresponding Turing machine classes. For example, *polynomial size* parity, ordinary, disjunctive, conjunctive and alternating branching programs define the newly discovered class $\oplus \mathcal{L}$ as well as such fudamental complexity classes like $\mathcal{L}$, $\mathcal{NL} = co\text{-}\mathcal{NL}$ and $\mathcal{P}$, respectively, which are strongly expected to be different. Unlike polynomial size *bounded width* Ω-branching programs, for all $\Omega \subseteq \mathbf{B}_2$ they define the same class coinciding with $\mathcal{NC}^1$. While all the classes defined by polynomial size-bounded and unbounded width Ω-branching programs are conjectured to be different most of the corresponding complexity classes defined by *quasipolynomial size* Ω-branching programs coincide for sure.

However, one of the most interesting and important tasks in complexity theory is to separate complexity classes such as $\mathcal{L}$, $\mathcal{NL}$ (= $co\text{-}\mathcal{NL}$) or $\mathcal{P}$ (or to prove their coincidence). Since we have described these complexity classes by means of polynomial size ordinary, disjunctive, conjunctive and alternating branching programs, respectively, superpolynomial lower bounds for such branching programs would essentially contribute to a separation of these classes. In cooperation with M.Krause and S.Waack we succeeded in separating ordinary, disjunctive, conjunctive and alternating read-once-only branching programs [KMW88] thus improving corresponding results for ordinary read-once-only branching programs [We84], [A&86], [KW87]. This is especially interesting since these read-once-only Ω-branching program complexity classes correspond to the nonuniform logarithmic space bounded eraser Turing machine classes $\mathcal{L}_e$, $\mathcal{NL}_e$, $co\text{-}\mathcal{NL}_e$ and $\mathcal{P}_e$. We obtain

$$
\begin{array}{ccccc}
 & & \mathcal{NL}_e & & \\
 & \overset{\subset}{\neq} & & \overset{\subset}{\neq} & \\
\mathcal{L}_e & & \cancel{A}\!\!\mid \quad \mid\!\!\cancel{\mid} & & \mathcal{P}_e = \mathcal{P} \quad . \\
 & \overset{\subset}{\neq} & & \overset{\subset}{\neq} & \\
 & & co\text{-}\mathcal{NL}_e & &
\end{array}
$$

Since up to now only $\mathcal{L}_e$ was separated we have carried out by $\mathcal{L}_e \overset{\subset}{\neq} \mathcal{NL}_e$, $\mathcal{L}_e \overset{\subset}{\neq} co\text{-}\mathcal{NL}_e$ and $\mathcal{NL}_e \overset{\subset}{\neq} \mathcal{P}_e$, $co\text{-}\mathcal{NL}_e \overset{\subset}{\neq} \mathcal{P}_e$ further steps in separating larger and larger complexity classes.

Finally, we are able to use the given branching program characterizations of various nonuniform complexity also for achieving p-projection completeness results. In particular, we have proved the p-projection completeness of a number of GRAPH-ACCESSIBILITY-PROBLEMS and certain restricted NETWORK-FLOW-PROBLEMS strengthening, unifying, and generalizing classical results of [Sa70], [CSV84]. Beside giving new insights into the capabilities of computations within certain complexity bounds varying the complexity of one and the same problem makes the "differences" of the corresponding complexity classes more evident.

This thesis contains the results of [Me86,1-2], [Me87,1-4], [Me88] and [KMW88], along with some additional material not yet published. It is organized as follows:

In the first chapter we introduce the concept of *branching programs* (Section 1.1) and review some previous results. In particular we are interested in the effect upon branching programs of tight constraints on width or on the input access. In Section 1.2 we review some of the previous results concerning *bounded width branching programs*. In particular we cite Barrington's result, which relates bounded width branching programs and certain depth restricted Boolean fan-in 2 circuits [Ba86]. The second important resource constraint for branching programs is that of restricting the number of accesses to the input variables on each path of computation. The most investigated case is that of *read-once-only branching programs* where each variable may be tested only once on each path. For read-once-only branching programs a number of even exponential lower bounds were obtained. A short review of these results is given in Section 1.3 .

In the second chapter we examine nondeterministic branching programs which have been introduced in [Me86,1]. Apart from giving the possibility to describe higher (nonuniform) complexity classes by means of certain branching programs the concept of nondeterminism overcomes some restrictions in the branching program model. Section 2.1 introduces nondeterministic branching programs and classifies them according to their ability to record nondeterministic choices. In Section 2.2 we compare the computational power of nondeterministic and 1-time-only-nondeterministic branching programs of polynomial size. In the final Section 2.3 the influence of bounding the width on the computational power of nondeterministic branching programs is studied.

Chapter 3 deals with Ω-branching programs, $\Omega \subseteq \mathbf{B}_2$, and their computational power. After classifying in Section 3.1 Ω-branching programs into the five types of ordinary, disjunctive $\{\vee\}$-branching programs, conjunctive $\{\wedge\}$-branching programs, parity $\{\oplus\}$-branching programs and alternating $\{\vee,\wedge\}$-branching programs we study the behavior of these types of Ω-branching programs under different resource constraints. At first, in Section 3.2, we let the size of such Ω-branching programs be polynomially bounded. In Section 3.3 we study polynomial size Ω-branching programs of bounded width. Section 3.4 examines quasipolynomial size Ω-branching programs, $\Omega \subseteq \mathbf{B}_2$. In order to separate larger complexity classes we investigate in the final Section 3.5 read-once-only Ω-branching programs, $\Omega \subseteq \mathbf{B}_2$. Indeed this approach proved to be quite successful since it is possible to separate the complexity classes which are related to polynomial size read-once-only Ω-branching programs, $\Omega \in \{\emptyset, \{\vee\}, \{\wedge\}, \{\vee,\wedge\}\}$.

In the appendix we consider some very restricted GRAPH-ACCESSIBILITY-PROBLEMS and prove their p-projection completeness in the complexity classes described in Chapters 1 to 3 by means of certain Ω-branching programs. For that purpose we give a number of Ω-branching programs which solve these GRAPH-ACCESSIBILITY-PROBLEMS.

A last technical remark should be made. Within each chapter all propositions, lemmas and corollaries are numbered continuously. In order to keep references within the text as short as possible the number of theorems is the same as that of the paragraph where the proof is given.

PRELIMINARIES

We assume that the reader is familiar with the two basic computation models of *Boolean circuits* and *Turing machines*.

Throughout this paper we treat only Boolean fan-in 2 circuits with one output node. Whenever we do not indicate the basis $\Omega \subseteq \mathbf{B}_2$ we think of $\{\vee, \wedge, (\neg)\}$-circuits. The *size* of a Boolean circuit C is the number of its gates. The *depth* is the length of a longest path from some input node to the output node. Since each circuit operates only on inputs of a given size we consider sequences of circuits and treat size and depth as functions of the input size. A sequence $\{C_n\}$ of circuits is said to be of *size* $S(n)$ or *depth* $D(n)$ if

$$Size(C_n) = O(S(n)) \quad \text{or} \quad Depth(C_n) = O(D(n)) \, ,$$

for each $n \in \mathbf{N}$. The sets of all languages $A \subseteq \{0,1\}^*$ accepted by sequences of circuits of size $S(n)$ or depth $D(n)$ will be denoted by $SIZE_{Cir}(S(n))$ or $DEPTH_{Cir}(D(n))$, respectively,

$$
SIZE_{Cir}(S(n)) = \{ \ A \subseteq \{0,1\}^* \mid \text{there is a sequence of circuits of size } S(n) \text{ which accepts } A \ \} \, ,
$$

$$
DEPTH_{Cir}(S(n)) = \{ \ A \subseteq \{0,1\}^* \mid \text{there is a sequence of circuits of depth } D(n) \text{ which accepts } A \ \} \, .
$$

Whenever size bound $S(n)$ and depth bound $D(n)$ are assumed to hold simultaneously we write

$$SIZE\text{-}DEPTH_{Cir}(S(n), D(n)) =$$

$$= \{\ A \subseteq \{0,1\}^* \mid \text{there is a sequence of}$$
circuits of size $O(S(n))$
and depth $O(D(n))$ which
accepts A } .

Of special interest in the field of parallel computation is the complexity class $\mathcal{NC}$ consisting of all languages which are acceptable by (sequences of) Boolean circuits of polynomial size and polylogarithmic depth

$$\mathcal{NC} = SIZE\text{-}DEPTH_{Cir}(n^{O(1)}, (\log n)^{O(1)}) .$$

Of special interest is the subclass $\mathcal{NC}^1$ of $\mathcal{NC}$,

$$\mathcal{NC}^1 = SIZE\text{-}DEPTH_{Cir}(n^{O(1)}, \log n) = DEPTH_{Cir}(\log n) ,$$

which coincides with the class of languages acceptable by (sequences of) polynomial length Boolean formulas.

A (deterministic, nondeterministic or alternating) Turing machine M is $f(n)$ *space-bounded* (resp. *time-bounded*) for some function f on the natural numbers if no computation of M on inputs of size n uses more than $f(n)$ tape squares (resp. if each computation of M has length at most $f(n)$). If f is bounded by a polynomial then M is *polynomial space* (resp. *polynomial time*) *bounded*. Related space complexity classes are:

$DSPACE(f(n)) = \{A \subseteq \{0,1\}^* \mid A$ can be accepted by a determinsitic $O(f(n))$ space bounded Turing machine} ,

$NSPACE(f(n)) = \{A \subseteq \{0,1\}^* \mid A$ can be accepted by a nondeterminsitic $O(f(n))$ space bounded Turing machine} ,

$ASPACE(f(n)) = \{A \subseteq \{0,1\}^* \mid A$ can be accepted by an alternating $O(f(n))$ space bounded Turing machine$\}$.

The time classes $DTIME(f(n))$, $NTIME(f(n))$ and $ATIME(f(n))$ are defined analogously.

For each complexity class K , let $co-K$ be the set of *complements* of languages of K

$$co-K = \{ A \subseteq \{0,1\}^* \mid \overline{A} \in K \} .$$

In particular we are interested in the following classes defined in terms of *logarithmic space bounded* or *polynomial time bounded* Turing machines:

$$L = DSPACE(log(n)) ,$$

$$NL = NSPACE(log(n)) ,$$

$$co-NL = co-NSPACE(log(n)) ,$$

$$AL = ASPACE(log(n)) , \text{ and}$$

$$P = DTIME(n^{O(1)}) .$$

For results comparing these complexity classes see any textbook on complexity theory (f.e [HU79]). Let us only mention the following two relations:

$$AL = P \text{ [CKS81]} \quad \text{and} \quad NL = co-NL \text{ [Im87,Sz87]}.$$

In order to relate complexity classes defined by Turing machines with circuit-based complexity classes one considers *nonuniform* Turing machine complexity classes [KL80]. If f is a function on the natural numbers and if K denotes any Turing machine complexity class then we define

$K/f(n) = \{A \subseteq \{0,1\}^* \mid$ there is an *advice* $\alpha : \mathbb{N} \longrightarrow \{0,1\}^*$ with $|\alpha(n)| \leq O(f(n))$ and a Turing machine operating within the resource constraints of K which accepts $w\#\alpha(|w|)$ iff $w \in A\}$,

and, for a set F of functions,

$$K/F = \bigcup_{f \in F} K/f .$$

If K is related to $s(n)$ space bounded Turing machine computations then the *nonuniform* counterpart $\mathcal{K}$ of K is defined by

$$\mathcal{K} = K / 2^{O(s(n))} .$$

If K is related to $t(n)$ time bounded Turing machine computations then the *nonuniform* counterpart $\mathcal{K}$ of K is defined by

$$\mathcal{K} = K / t(n) .$$

In particular we are interested in the following nonuniform complexity classes

$$\mathcal{L} = L / n^{O(1)} ,$$
$$\mathcal{NL} = NL / n^{O(1)} , \text{ and}$$
$$\mathcal{AL} = AL / n^{O(1)} ,$$

which coincides with

$$\mathcal{P} = P / n^{O(1)} .$$

It is well-known that the nonuniform Turing machine class $\mathcal{P}$ consists of all languages recognizable by means of (sequences of) polynomial size Boolean circuits.

$$\mathcal{P} = SIZE_{Cir}(n^{O(1)}) .$$

CHAPTER 1

BRANCHING PROGRAMS AND THEIR COMPUTATIONAL POWER

INTRODUCTION

A fundamental issue of complexity theory is to estimate the relative efficiency of different models of computation. One of the most important models for investigating the computational complexity of Boolean functions is that of branching programs which generalize the concept of decision trees to decision graphs. The settings of n input variables determine a flow of control through a directed network of processors, as each processor triggers one of two successors in dependence on the value of one of the input bits. Branching programs were first studied by Lee [Le59] (under the name 'binary decision programs') as an alternative to Boolean circuits in the description of switching problems. A first systematical study of branching programs can be found in the Master's thesis of Masek [Ma76].

The present chapter introduces the concept of branching programs and reviews some previous results.

In Section 1.1 we define branching programs and present a full proof of the relations between branching program size and space complexity of nonuniform Turing machines first proved by Cobham [Co66], Budach [Bu82], Pudlák and Žák [PŽ83].

Although it is known that the computation of most Boolean

functions requires exponential size branching programs the largest known lower bound for an explicitly given Boolean function is an $\Omega(n^2/(\log n)^2)$ bound of Nechiporuk [Ne66]. In order to gain more insight into the problem of proving lower bounds one has successfully investigated more restricted branching programs.

One of such restricted models is that of bounded width branching programs. It was introduced by Borodin, Dolev, Fich and Paul [BDFP83]. In Section 1.2 we review some of the previous results concerning such bounded width branching programs. In particular we cite Barrington's result [Ba86], which relates bounded width branching programs and certain depth restricted Boolean fan-in 2 circuits.

A second important resource constraint for branching programs is that of restricting the number of tests of the input variables on each path of computation. The best investigated case is that of read-once-only branching programs. In this model each variable may be tested on each path only once. For read-once-only branching programs a number of even exponential lower bounds were obtained, e.g. [We84], [Ža84], [A&86], [KW87]. A short review of these results is given in Section 1.3 .

1.1. Branching Programs

A *branching program* (*BP*) is a directed acyclic graph where each node has outdegree 2 or 0 . Nodes with outdegree 0 are called *sinks* and labelled by Boolean constants. The remaining nodes are labelled by Boolean variables taken from a set $X = \{x_1,...,x_n\}$. There is a distinguished node, called the *source*, which has indegree 0 . A branching program *computes* an n-argument Boolean function $f \in \mathbb{B}_n$ as follows: Starting at the source, the value of the variable labelling the current node is tested. If this is 0 (1) , the next node which will be tested is the left (right) successor to current node. The branching program *computes* f if, for the input $w \in \{0,1\}^n$, the path traced under w halts at a sink labelled by $f(w)$. Without loss of generality we may assume that a branching program has exactly two sinks, one 0-sink and one 1-sink.

Via the usual correspondence between binary languages $A \subseteq \{0,1\}^*$ and sequences $\{f_n\}$ of Boolean functions $f_n \in \mathbb{B}_n$, namely

$$w \in A \quad \text{iff} \quad f_{|w|}(w) = 1 ,$$

a sequence $\{P_n\}$ of branching programs is said to *accept a language* $A \subseteq \{0,1\}^*$ if, for all $n \in \mathbb{N}$, P_n computes the characteristic function $\chi_{A^n}(w)$ of the n-th *restriction* A^n of A ,

$$A^n = A \cap \{0,1\}^n .$$

The *length* (sometimes also called the *depth*) of a branching program P is the length of a longest path in P. However, the most important complexity measure of a branching program is its size. The *size* of a branching program P is equal to the number of nodes of P

$$Size(P) \;=\; \# \, P \, .$$

A sequence $\{P_n\}$ of branching programs accepting a set $A \subseteq \{0,1\}^*$ is said to be of *size* $S(n)$, if

$$Size(P_n) \;=\; O(S(n))$$

for all $n \in \mathbb{N}$. If we investigate a branching program belonging to a sequence of branching programs of size $S(n)$ we will speak, for short, of a branching program of size $S(n)$, or of a $S(n)$ size branching program. The set of all languages $A \subseteq \{0,1\}^*$ acceptable by sequences of branching programs of size $S(n)$ will be denoted by $SIZE_{BP}(S(n))$,

$$SIZE_{BP}(S(n)) = \{ \, A \subseteq \{0,1\}^* \;\mid\; \text{there is a sequence of branching programs of size } S(n) \text{ accepting } A \, \} \, .$$

The class of languages acceptable by (sequences of) *polynomial size* branching programs is of special interest. We denote this class by $\mathcal{P}_{BP}$,

$$\mathcal{P}_{BP} \;=\; SIZE_{BP}(n^{O(1)}) \, .$$

The following relation between branching program size and space complexity of nonuniform Turing machines is well-known and nowadays a standard result. It goes back to Cobham [Co66] and was first proved in the nonuniform settings by Pudlák and Žák [PŽ83]. Since we will generalize this relation to more powerful branching programs and Turing machines in the course of this paper we give a full prove of this relation.

THEOREM 1.1.

Let $s(n) = \Omega(\log n)$. Sequences of branching programs of size $2^{O(s(n))}$ and $s(n)$ space-bounded nonuniform Turing machines are of the same computational power. I.e.

$$SIZE_{BP}(2^{O(s(n))}) = SPACE(s(n)) \, / \, 2^{O(s(n))} \, .$$

PROOF.

Let $s(n) = \Omega(\log n)$, let $A \in SIZE_{BP}(2^{O(s(n))})$, and let $\{P_n\}$ be a sequence of branching programs of size $S'(n)$, $S'(n) = 2^{O(s(n))}$, accepting A . We construct a $s(n)$ space-bounded nonuniform Turing machine simulating the computations of sequence $\{P_n\}$ by using, for each $n \in \mathbb{N}$, a coding of P_n as advice for inputs of length n : Let ν be an enumeration of the vertex set V of P_n which assigns 0 to the source v_0 and $(S'(n) - 1)$ to the 1-sink of P_n . Define the advice $\alpha(n)$ by

$$\alpha(n) = \#\#\nu(v_0)\# b(v_0)\#\nu(l(v_0))\#\nu(r(v_0))...$$

$$..\#\#\nu(v_{S'(n)-2})\# b(v_{S'(n)-2})\#\nu(l(v_{S'(n)-2}))\#\nu(r(v_{S'(n)-2}))..$$

$$..\#\#\nu(v_{S'(n)-1})\# b(v_{S'(n)-1})\#\#$$

with

- $b(v)$ denotes the Boolean constant or the binary coding $bin(i)$ of the Boolean variable x_i the node v is labelled with, and

- $l(v)$ and $r(v)$ denote the left and the right successor node of v , respectively. If v is a sink, then $l(v)$ and $r(v)$ are blank symbols.

Obviously,

$$|\alpha(n)| \leq S'(n) \cdot (3 \cdot log \ S'(n) + log \ n + 5) \ .$$

Since $S'(n) = 2^{O(s(n))}$ it holds

$$|\alpha(n)| = 2^{O(s(n))} \ .$$

Now we describe a Turing machine M which accepts $w\#\alpha(n)$, $w \in \{0,1\}^n$, iff $w \in A$: M starts its computation by writing the string $\#\#0\#$ on the working tape. Until it finds $\#\#\mu\#\delta\#\#$ as the content of the working tape with $\delta \in \{0,1\}$ (the label of the sink reached in the simulated branching program) and $\mu = \nu(v)$ (the number of that sink) M repeats the following procedure: it looks for the substring $\#\#\mu\#$ on the input tape. After finding it, M copies the missing information of the succeeding subword $(\#\#\mu)\#u\#\mu'\#\mu''\#\#$. If $u = bin(i)$, then M will read the bit x_i on the i-th position of the input tape. Finally, M replaces μ by μ' (resp. μ'') if $x_i = 0$ (resp. $x_i = 1$) and erases the rest of the working tape inscription. M accepts iff it finds $\#\#\nu(v_{S'(n)-1})\#1\#\#$, i.e. iff the path traced under w ends up in the 1-sink.

Obviously, M is $O(log \ S'(n)) = O(s(n))$ space-bounded.

Now, let M be a $s(n)$ space-bounded Turing machine computing a set A nonuniformly. Fix an input length n , and let $\alpha(n)$ be the advice of length $|\alpha(n)| = 2^{O(s(n))}$. We design a branching program P_n which simulates M on inputs of length n from the set of configurations of M . The configurations of M are the 4-tuples (q,i,u,j) , where q is a state of the finite control, i with $0 \leq i \leq n + |\alpha(n)| + 1$ is the position of the input head on the input string, the word u with $|u| = O(log \ S(n))$ is the content of the working tape, and j is the position of the head on the working tape.

The machine has a unique initial configuration

$$C_0 = (q_0, 0, \#^{O(s(n))}, 1)$$

where q_0 is the unique initial state. W.l.o.g. we can assume that M has a unique accepting configuration

$$C_a = (q_a, 0, \#^{O(s(n))}, 1)$$

too, where q_a is the unique accepting state of the finite control of M. Now we modify M such that it will never halt, and that, if configuration C_a is entered, then M will cycle forever in C_a. Since M was assumed to be $s(n)$ space-bounded it has at most $m = 2^{O(s(n))}$ different configurations. Obviously, this number m can be taken as an upper bound for the time for M to accept. I.e. if M accepts an input $w\#\alpha(n)$, $|w| = n$, then M will be in configuration C_a at step m.

Now we are ready for designing a branching program P_n computing F_n. The vertices of P_n are pairs (C,t) where $C = (q,i,u,j)$ is a configuration of M with $i \leq n + |\alpha(n)| + 1$ and $0 \leq t \leq m$. The source of P_n is $(C_0,0)$. If (C,t), $C = (q,i,u,j)$ is a node of P_n, then we will assign the following Boolean variables and Boolean constants, respectively, to it:

$$b(C,t) = \begin{cases} 1 & \text{if } (C,t) = (C_a,m) \text{ ;} \\ x_i & \text{if } C = (q,i,u,j) \text{ and } t < m \text{ ;} \\ 0 & \text{otherwise.} \end{cases}$$

Depending on the value of the i-th digit $x_i = b(C,t)$ of the input string $w\#\alpha(n)$, M reaches the configuration

$$C(x_i) = (q(x_i), i(x_i), u(x_i), j(x_i)), \quad x_i \in \{0,1\}$$

in one move. We add $(C(x_i),t+1)$ to the vertices of P_n and define

$$l(C,t) \;=\; (C(x_i), t+1) \qquad \text{iff} \quad x_i = 0 \;,$$

or

$$r(C,t) \;=\; (C(x_i), t+1) \qquad \text{iff} \quad x_i = 0 \;.$$

Since the advice $\alpha(n)$ depends only on n and not on the first part w, $|w| = n$, of the input string $w\#\alpha(n)$ we can "hardwire" the x_i, $i > n$. I.e. we successively identify all vertices (C,t) labelled x_i, $i > n$ with their left (if $x_i = 0$) or right (if $x_i = 1$) successor nodes $(C(x_i), t+1)$, respectively. Finally, we delete all nodes (C,t) of P_n which are not reachable from the source $(C_0, 0)$.

Obviously, P_n is acyclic because all edges of P_n lead to vertices with a strictly higher second component t. Since $s(n) = \Omega(\log n)$ the size of P_n is bounded by $m^2 = O(s(n))$. Finally, P_n is a branching program computing $A^n = A \cap \{0,1\}^n$, since for each input w, $|w| = n$, there is a path from the source $(C_0, 0)$ to the accepting node (C_a, m) iff M accepts $w\#\alpha(n)$, i.e. iff $w \in A^n \subseteq A$. ∎

In particular we obtain:

COROLLARY 1.

Polynomial size branching programs and logarithmic space-bounded nonuniform Turing machines are of the same computational power. I.e.

$$\mathcal{P}_{BP} \;=\; \mathcal{L} \;. \quad \blacksquare$$

1.2. Bounded Width Branching Programs

Let us call a branching program P *synchronous* (or *levelled*) if for each node v of P all paths from the source to v are of the same length $d(v)$. It is easy to see that we can make an arbitrary branching program synchronous by adding dummy nodes, possibly squaring the size but keeping the length the same. *Level* j of such a synchronous branching program P consists of all nodes v with $d(v) = j$. Hence, a branching program of length l has $l+1$ levels. The *width* $w(j)$ of a level j is the number of nodes v with $d(v) = j$. The *width* of P is the maximum, over $0 \leq j \leq l$, of $w(j)$

$$Width(P) \quad = \quad \max_{0 \leq j \leq l} \quad \{w(j)\} \ .$$

A sequence $\{P_n\}$ of synchronous branching programs is said to be of width $W(n)$ if

$$Width(P_n) \quad = \quad O(W(n))$$

for all $n \in \mathbb{N}$. If we investigate a branching program P belonging to a sequence of width $W(n)$ branching programs we will speak, for short, of P as of a width $W(n)$ branching program. The set of all languages $A \subseteq \{0,1\}^*$ acceptable by sequences of branching programs of size $S(n)$ and width $W(n)$ will be denoted by $SIZE\text{-}WIDTH_{BP}(S(n), W(n))$,

$$SIZE\text{-}WIDTH_{BP}(S(n), W(n)) =$$

$= \{ A \subseteq \{0,1\}^* \mid$ there is a sequence of branching programs of size $S(n)$ and width $W(n)$ which accepts $A \}$.

A branching program is said to be in *normal form* if, with

exception of the last level, every level contains the same number of nodes and if all nodes of a level are labelled by the same Boolean variable x_i , $1 \leq i \leq n$. The last level, level l , consists of a 1-sink and a 0-sink. We assume the source to be the left most node of level 0 . A straightforward argument shows that every synchronous branching program can be converted to a normal form branching program which accepts the same set at the cost of doubling the width and multiplying the length by the minimum of the width and the number n of input variables.

In the case of constant width $W(n) = O(1)$ we speak of (sequences of) *bounded width branching programs*. Obviously, the size of a bounded width branching program equals its length, to within a constant factor (namely the width). Moreover, since we are only interested in complexity results to within constant (polynomial) factors we can assume bounded width branching programs to be in normal form. It is well-known that all Boolean functions $\omega \in \mathbb{B}_n$ can be computed already by width-2 branching programs. However the length of such programs may increase up to exponential size, e.g. width-2 branching programs for the *majority* function cannot have polynomial size [Ya83].

A very interesting branching program complexity class is the class $\mathcal{P}_{bw-BP}$ of all languages acceptable by (sequences of) bounded width branching programs of polynomial size

$$\mathcal{P}_{bw\ BP} \quad = \quad SIZE\text{-}WIDTH_{BP}(n^{O(1)},\ 1)\ .$$

Before we present a characterization of $\mathcal{P}_{bw\ BP}$, recently given by Barrington [Ba86], we will briefly review the history of lower bounds of width-restricted branching programs. The bounded width problem was introduced in [BDFP83]. We have already mentioned the result of Yao [Ya83] stating that the *majority* function cannot be computed by polynomial size branching programs of width 2. For $k \geq 3$ no large lower bounds on the width-k branching program complexity of explicitly defined Boolean functions are known. By arguments from Ramsay theory it was proved that the *majority* function cannot be computed by bounded width and linear size branching programs [CFL83]. Further linear and superlinear length (and, consequently, size) lower bounds for arbitrary bounded width branching programs were proved in [Pu84]. In [A&86] it was shown that almost all symmetric functions cannot be computed by branching programs of polylogarithmic width and size o($n(log\ n)/log\ log\ n$). All these results were motivated by (and appeared to support) the conjecture that the majority function cannot be calculated by polynomial size branching programs of bounded width [BDFP83]. However, this conjecture has been refuted by Barrington [Ba86]. Since *majority* is a symmetric function and since each symmetric function can be computed in NC^1 [MP75] the following theorem implies the existence of polynomial size bounded width branching programs for *majority*.

THEOREM 1.2 ([Ba86]).

Let $S(n) = \Omega(n^{O(1)})$. *Bounded width branching programs of size* $S(n)$ *and Boolean (fan-in 2) circuits of depth* $log\ S(n)$ *are of the same computational power. I.e.*

$$SIZE\text{-}WIDTH_{BP}(S(n),\ 1) = DEPTH(log\ S(n)) . \blacksquare$$

In particular we obtain

COROLLARY 2.

Polynomial size branching programs of bounded width and Boolean circuits of logarithmic depth are of the same computational power. I.e.

$$\mathcal{P}_{bw-BP} = \mathcal{N}\mathcal{C}^1 . \blacksquare$$

It should be mentioned that Barrington has proved that each language acceptable by (a sequence of) fan-in 2 Boolean circuits of depth $D(n)$ may already be accepted by (a sequence of) width 5 branching programs of length at most $D(n) \cdot 4^{D(n)}$. (In more detail, Barrington has proved this result for the special type of *permutation branching programs*. However, a width w permutation branching program of length l may be simulated by an (ordinary) branching program of width w and length $w \cdot l$ [We87].) The simulation of bounded width branching programs of size $S(n)$ by Boolean (fan-in 2) circuits of depth $O(log\ S(n))$ can be obtained from the nonuniform version of an argument which essentially appears in [Sa70] and is given explicitly in [LF77].

1.3. READ-ONCE-ONLY BRANCHING PROGRAMS

Branching programs where each variable is tested (read) on each computation path at most once are called *read-once-only branching programs (BP1)*. This type of computation model was introduced by Masek [Ma76]. The corresponding Turing machine model is the nonuniform *eraser Turing machine*. That are Turing machines which erase each input bit after having read it. Similar to Theorem 1.1 one obtains

THEOREM 1.3 ([A&86]).
Let $s(n) = \Omega(log\ n)$. Sequences of read-once-only branching programs of size $2^{O(s(n))}$ and $s(n)$ space-bounded nonuniform eraser Turing machines are of the same computational power.

I.e.

$$SIZE_{BP1}(2^{O(s(n))}) = DSPACE_e(s(n))/2^{O(s(n))} ,$$

where $DSPACE_e(f(n))$ *denotes the class of languages recognizable by* $f(n)$ *space-bounded eraser Turing machines.* ∎

If $\mathcal{F}_{BP1}$ denotes the set of languages acceptable by (sequences of) polynomial size read-once-only branching programs and if L_e and $\mathcal{L}_e$ denote the complexity classes related to logarithmic space bounded uniform and nonuniform eraser Turing machines, respectively, then Theorem 1.3 yields

COROLLARY 3.

Polynomial size read-once-only branching programs and logarithmic space-bounded eraser Turing machines are of the same computational power. I.e.

$$\mathcal{F}_{BP1} = \mathcal{L}_e . ∎$$

Due to Theorem 1.3 , lower (and upper) bounds on the read-once-only branching program complexity yield lower (and upper) bounds on the space complexity of eraser Turing machines. Interest in read-once-only branching programs stems from the fact that, in contrast to the situation in general branching programs, even exponential lower bounds are known for some explicitly defined functions. The first nearly exponential lower bound for read-once-only branching programs was given by Wegener [We84] and Žak [Ža84] for certain clique functions. Such a *clique function* $cl_{n,m}$ decides for a given undirected n-node graph G whether it contains a m-clique (that are m vertices any two of which are adjacent). If $G = G(x)$ is represented by the adjacency matrix $x = ((x_{ij}))$, $1 \le i < j \le n$, with $x_{ij} = 1$ if nodes i and j of G are connected by an edge, then it holds $cl_{n,m}(x) = 1$ iff $G = G(x)$ contains an m-clique. Wegener's proof of the nearly exponential lower bound for the size of read-once-only branching programs computing the clique function $cl_{n,n/2}$ is based on the following idea. He shows that certain computation paths in a branching program for $cl_{n,n/2}$, whose lengths are at most d, cannot end in a sink and cannot be merged with other computation paths. Hence, each read-once-only branching program for $cl_{n,n/2}$ contains at its top a complete binary tree of depth d and therefore at least $2^d - 1$ inner nodes.

In the meantime this idea has found many successful applications. Other nearly exponential lower bounds were obtained by Dunne [Du85] for the logical permanent and Hamiltonian circuit functions and by Krause [Kr86] for deciding certain algebraic properties of Boolean matrices and for some interesting subgraph problems.

A first proper exponential lower bound was obtained by Ajtai, Babai, Hajnal, Kolmós, Pudlák, Rödl, Szemerédi and Turán [A&86] for the Boolean function $\oplus cl_{n,3} = \oplus cl_{n,3}(x)$, $x = (x_{ij})$ $(1 \le i < j \le n)$ which decides whether a given undirected n-node graph $G = G(x)$ contains an odd number of triangles (3-cliques).

LEMMA 4 ([A&86]).

Each read-once-only branching program which computes $\oplus cl_{n,3}$ is of size $2^{\Omega(N)}$, $N = \binom{n}{2}$. ∎

Further exponential lower bounds for the read-once-only branching program complexity with more elegant proofs were obtained by Kriegel and Waack [KW87] for Dyck languages and by Krause [Kr87] for certain graph problems.

Chapter 2

NONDETERMINISTIC BRANCHING PROGRAMS

Introduction

The following chapter is devoted to the study of nondeterministic branching programs which have been introduced in
[Me86,1]. Beside giving the possibility to describe higher
(nonuniform) complexity classes by means of the circuit based
computational model of nondeterministic branching programs the
concept of nondeterminism overcomes some restrictions in the
branching program model.

In Section 2.1 we introduce nondeterministic branching programs and classify them on the basis of their ability to record
nondeterministic choices. We show that already 2-times-only-
nondeterministic branching programs are as powerful as (unrestricted) nondeterministic branching programs (Theorem 2.1),
whereas 1-time-only-nondeterministic branching programs, which
are unable to record nondeterministic choices, seem to be much
less powerful.

In Section 2.2 we describe the computational power of polynomial size 1-time-only-nondeterministic branching programs
(Paragraph 2.2.1) and that of (unrestricted) nondeterministic
branching programs of polynomial size (Paragraph 2.2.2). In detail, we obtain the following results:

(i) 1-time-only-nondeterministic branching programs of polynomial size are as powerful as nondeterministic logarithmic
space-bounded nonuniform Turing machines (Theorem 2.2.1)
[Me86,1], and

(ii) nondeterministic branching programs of polynomial size are
as powerful as nondeterministic polynomial time bounded
nonuniform Turing machines (Theorem 2.2.2) [Me86,2].

Finally, in Section 2.3 we examine the influence of bounding the width on the computational power of nondeterministic
branching programs. We prove that 1-time-only-nondeterministic
polynomial size branching programs of bounded width are no more
powerful than deterministic ones (Theorem 2.3.1) [Me87,3],
while k-times-only-nondeterministic polynomial size branching
programs of bounded width, $k \geq 2$, are as powerful as nondeterministic polynomial size branching programs without any width
restrictions (Theorem 2.3.2) [Me87,3].

2.1. NONDETERMINISTIC BRANCHING PROGRAMS AND THEIR CLASSIFICATION

A branching program P is called a *nondeterministic branching program accepting* a set $A^n \subseteq \{0,1\}^n$ if there is a
function

$$h : \{0,1\}^{n+m} \longrightarrow \{0,1\} \quad , \quad m \geq 0$$

with

$$\chi_{A^n}(x_1,...,x_n) = \bigvee_{x_j \in \{0,1\}, \, j>n} h(x_1,...,x_n,x_{n+1},...,x_{n+m}),$$

and if P is a branching program computing h [Me86,1]. In
contrast to ordinary (deterministic) branching programs nonde-

terministic branching programs possibly provide different computation paths for one input $(x_1,...,x_n) \in \{0,1\}^n$ depending on the settings of the *nondeterministic variables* x_j , $j > n$. P accepts if at least one of these computation paths is an accepting one. However, the number of nondeterministic choices is bounded by the size of the whole program.

Examining nondeterministic branching programs one realizes that the ability of a nondeterministic branching program to record its nondeterministic choices will greatly influence its computational power. This ability can be quantified by the number of different accesses to the nondeterministic variables during a computation. A nondeterministic branching program P is called *k-times-only-nondeterminstic* if each of the nondeterministic variables x_j , $j > n$, is tested at most k times on any path of P from the source to a sink . Obviously, in k-times-only-nondeterministic branching programs the chosen values of the nondeterministic variables can be referred to at most k times during any computation. Particularly, 1-time-only-nondeterministic branching programs are unable to record their nondeterministic choices and to use this information repeatedly.

Interestingly, already 2-times-only-nondeterministic branching programs are as powerful as unrestricted nondeterministic branching programs as can be seen from the following theorem.

Two nondeterministic branching programs are said to be *computationally equivalent* if they accept the same set and if their sizes coincide, to within a constant factor.

THEOREM 2.1.

For each nondeterministic branching program there is a computationally equivalent 2-times-only-nondeterministic branching program.

PROOF.

Let P be a nondeterministic branching program with the nondeterministic variables x_j, $n < j \le n+m$. We can first obtain a 1-time-only-nondeterministic branching program P'' by replacing the nondeterministic variables x_j, $n < j \le m$, assigned to nodes v of P by new nondeterministic variables $x_{j,v}$. We will get a nondeterministic branching program P' accepting the same set as P does, if we check, in addition to an accepting computation of P'', the Boolean equivalences

$$x_j \leftrightarrow x_{j,v_t}, \quad 1 \le t \le r_j,$$

for all nodes $v \in \{v_1, ..., v_{r_j}\}$ of P labelled by x_j before accepting an input. Since the following nondeterministic branching programs P_j, $n < j \le n+m$, perform this job,

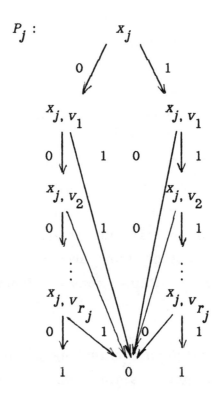

we can obtain P' by identifying the 1-sink of P'' with the source of P_{n+1}, and the 1-sink of P_j, $n < j < n+m$, with the source of P_{j+1}.

Since P' is 2-times-only-nondeterministic and of size

$$Size(P') \leq 4 \cdot Size(P)$$

we are done. ∎

Due to Theorem 2.1 we can divide nondeterministic branching programs into two classes: the class of 1-time-only-nondeterministic branching programs and the class of all remaining nondeterministic branching programs. Later, in Section 2.2 we reveal differences in the computational power of these two types of nondeterministic branching programs. They also differently respond to bounding their width as will be shown in Section 2.3.

2.2. The Computational Power of Nondeterministic Branching Programs of Polynomial Size

In this section we examine the computational power of nondeterministic branching programs of polynomial size. Due to the classification of nondeterministic branching programs given in Section 2.1 we establish a significant difference in the computational power between polynomial size 1-time-only-nondeterministic branching programs and the remaining nondeterministic branching programs of polynomial size.

2.2.1. 1-TIME-ONLY-NONDETERMINISTIC BRANCHING PROGRAMS OF POLYNOMIAL SIZE

In the following we prove that the class $\mathcal{P}_{n_1BP}$ of languages acceptable by sequences of 1-time-only-nondeterministic branching programs of polynomial size coincides with the class $\mathcal{NL}$ of languages recognizable by nondeterministic logarithmic space-bounded nonuniform Turing machines. This result generalizes the analogously deterministic statement of Theorem 1.1 in Chapter 1 . It was first published in [Me86,1].

THEOREM 2.2.1.

Sequences of 1-time-only-nondeterministic branching programs of polynomial size and nondeterministic logarithmic space-bounded nonuniform Turing machines are of the same computational power, i.e.

$$\mathcal{P}_{n_1}BP = \mathcal{NL} .$$

PROOF.

Analogous to the deterministic case described in Theorem 1.1 in Chapter 1 we obtain $\mathcal{P}_{n_1}BP \subseteq \mathcal{NL}$ by allowing the advice $\alpha(n)$ to encode the 1-time-only-nondeterministic polynomial size branching program P_n for inputs of length n, and taking an almost analogous but nondeterministically working logarithmic space bounded Turing machine M which guesses the values of x_j if $j > n$. After having guessed such a nondeterministic value M proceeds in the same way as with the deterministic values of the input variables. Since P_n was assumed to be 1-time-only-nondeterministic M can forget the guessed value because every input x_j, $j > n$, occurs at most once on each computation path. This fact together with the polynomial size of P_n, which implies the polynomial length of the advice $\alpha(n)$, keeps M logarithmic space-bounded, too.

Now let $A \in \mathcal{NL} = NL/n^{O(1)}$, and let M be a nondeterministic logarithmic space-bounded nonuniform Turing machine, recognizing A. From [Pa78] it is well known that we can assume M to have at most two nondeterministic choices in every step. As in the proof of the deterministic case (Theorem 1.1, Chapter 1) let $\alpha(n)$ be the advice of polynomial length in the input length n and let C_0 and C_a be the uniquely determined initial and accepting configurations. Once more let us assume that M is modified in such a way that M cycles in the accepting configuration C_a forever if C_a is reached in the course of computation.

The vertices of the desired 1-time-only-nondeterministic branching program P_n, $n \in \mathbb{N}$, computing $A^n = A \cap \{0,1\}^n$, are chosen from the sets

$\{(C, t) : \quad C$ is a configuration of M, $0 \le t \le m\}$

and

$\{((C, C'), t) : \quad C, C'$ are configurations of M, $0 \le t \le m\}$,

where m is the number of distinct configurations which is polynomial in n. Thus the polynomial size of P_n is guaranteed.

P_n can be constructed as follows. Take $(C_0, 0)$ as the source of P_n. If (C, t), $C = (q, i, u, j)$ is a node of P_n, then assign the following Boolean variables orBoolean constants to it:

$$b(C, t) \quad = \quad \begin{cases} 1 & \text{if} \quad (C, t) = (C_a, m), \\ x_i & \text{if} \quad C = (q, i, u, j) \text{ and } t < m, \\ 0 & \text{otherwise}. \end{cases}$$

Now, depending on the value of the i-th digit x_i of the input word the configuration $C = (q, i, u, j)$ of M has at most two successor configurations

$$C(x_i) \quad \text{and} \quad C'(x_i), \quad x_i \in \{0, 1\}.$$

For $x_i = 0$ we add $(C(0), t+1)$ and $((C(0), C'(0)), t)$ to the vertices of P_n and define

$$l(C, t) \quad = \quad \begin{cases} (C(0), t+1) & \text{if} \quad C(0) = C'(0), \\ ((C(0), C'(0)), t) & \text{otherwise}. \end{cases}$$

For $x_i = 1$ we add $(C(1), t)$ and $((C(1), C'(1)), t)$ to the vertices of P_n and define

$$r(C, t) \quad = \quad \begin{cases} (C(1), t+1) & \text{if} \quad C(1) = C'(1), \\ ((C(1), C'(1)), t) & \text{otherwise}. \end{cases}$$

If $((C,C'),t)$ is a node of P_n, then we assign the variables $x_{C,C',t}$

$$b((C,C'),t) = x_{C,C',t},$$

add $(C,t+1)$ and $(C',t+1)$ to the vertices of P_n and define

$$l((C,C'),t) = (C,t+1) \quad \text{and} \quad r((C,C'),t) = (C',t+1).$$

Note that the variables $x_{C,C',t}$, $t < m$, stand for the non-deterministic choices of P_n. They are used for guessing one of the possible successor configurations C and C'. After hardwireing x_j, $j > n$, with respect to the advice $\alpha(n)$, and deleting all vertices which are not accessible from the source we obtain the desired P_n.

Obviously, P_n is acyclic. Further, it can easily be checked that P_n is a nondeterministic branching program for A^n. If $w \in A^n$, i.e. if there is an accepting computation of M for the input w, then there is a path from $(C_0,0)$ to (C_a,m) in P_n. On the other hand, $w \notin A^n$ implies that there is no accepting computation of M. Hence, starting in node $(C_0,0)$ of P_n, the accepting node (C_a,m) will never be reached. But this means exactly

$$\chi_{A^n}(w) = \bigvee_{y \in \{0,1\}^{\#b(P_n)-(n-2)}} h(w,y)$$

where $h(w,y)$ denotes the function computed by P_n.

Finally, assigning the number t of steps to the nondeter-ministically guessed variables $x_{C,C',t}$ will guarantee that these variables are tested at most once on each path of P_n, i.e. P_n is 1-time-only-nondeter- ministic and $A \in \mathcal{P}_{n_1 BP}$. $\blacksquare$

In order to make things easy we have constructed a relati-vely inefficient 1-time-only-nondeterministic branching program. In general, all vertices (C, t) of P_n consisting of a configuration C with only one successor configuration that is independent of the input can be omitted.

2.2.2. NONDETERMINISTIC BRANCHING PROGRAMS OF POLYNOMIAL SIZE

The computational power of nondeterministic branching programs increases enormously if, in the course of computation, repeated access to the same nondeterministic variable is allowed. Under this assumption the class $\mathcal{P}_{nBP}$ of languages acceptable by (sequences of) polynomial size nondeterministic branching programs coincides with the class $\mathcal{NP}$ of languages computable by nondeterministic polynomial time-bounded nonuniform Turing machines. This result was first published in [Me86,2] and can be found in [Me87,1], too.

THEOREM 2.2.2.

Sequences of polynomial size nondeterministic branching programs and nondeterministic polynomial time-bounded nonuniform Turing machines are of the same computational power, i.e.

$$\mathcal{P}_{nBP} = \mathcal{NP} .$$

The **PROOF** of Theorem 2.2.2 is a consequence of the following three lemmas.

LEMMA 1.

Each sequence of polynomial size nondeterministic branching programs can be simulated by a nondeterministic polynomial time-bounded nonuniform Turing machine, i.e.

$$\mathscr{P}_{nBP} \subseteq \mathscr{NP} .$$

PROOF.

Let $A \in \mathscr{P}_{nBP}$, and let P_n be a nondeterministic branching program accepting the n-th restriction $A^n = A \cap \{0,1\}^n$ of A. Once more we encode P_n by the advice $\alpha(n)$. A nondeterministic Turing machine M simulating P_n can be constructed analogously as in the proof of Theorem 2.2.1 . But unlike there we equip the nondeterministic Turing machine M with an additional working tape for recording the nondeterministic choices of P_n . On this additional working tape M writes the guessed values $B(j)$ of the nondeterministic variables x_j , $j > n$, e.g. as $\# bin(j) \# B(j) \#$. If M reaches a node v of P_n which is labelled by a nondeterministic variable x_j , $j > n$, then M looks for the substring $\# bin(j) \# B(j) \#$ on this working tape. If it finds this substring then M proceeds with the value $B(j)$. Otherwise it guesses a value $B(j)$, adds the string $\# bin(j) \# B(j) \#$ to the end of the additional tape, and proceeds with $B(j)$. Since the size of P_n was assumed to be polynomial in n we have $|\alpha(n)| \le p(n)$ for a polynomial p , and, consequently, $j \le p(n)$. Hence the needs of space and time of M remain polynomial. $\square$

Instead of proving $\mathscr{NP} \subseteq \mathscr{P}_{nBP}$ by encoding nondeterministic Turing machine computations by nondeterministic branching programs directly we apply the usual techniques for encoding (polynomial time) Turing machine computations by (polynomial size) Boolean circuits [La75] . These techniques and the well known characterization of NP by means of polynomially bounded quantifiers allow the simulation of nondeterministic Turing machines by nondeterministic circuits.

Before giving such a simulation let us recall the mentioned characterization of *NP* by means of polynomially bounded quantifiers given in [Wr76], [St77] : A set $A \subseteq \{0,1\}^*$ can be recognized by a nondeterministic polynomial time-bounded Turing machine if and only if there is a polynomial p and a set A' recognizable by a deterministic polynomial time-bounded Turing machine with

$$A = \{ w \mid \exists y \in \{0,1\}^{p(|w|)} : w\#y \in A' \} .$$

Nondeterministic circuits introduced in [SV85] are defined as follows: A Boolean $\{\vee,\wedge,\neg\}$-circuit C whose input nodes are labelled by Boolean constants and Boolean literals over the set $\{x_1,...,x_n,y_1,...,y_m\}$ is said to compute a set $A^n \subseteq \{0,1\}^n$ *nondeterministically* if exactly for all $w \in A^n$ there is an assignment of Boolean constants to the *nondeterministic variables* $y_1,...,y_m$ such that C outputs 1 ,

$$\chi_{A^n}(w) = \bigvee_{y \in \{0,1\}^m} C(w,y) .$$

$\mathscr{P}_{nCir}$ denotes the class of languages computable by (sequences of) polynomial size nondeterministic circuits.

Now we proceed in the following two steps: first we prove $\mathscr{NP} \subseteq \mathscr{P}_{nCir}$ (Lemma 2) and then we show $\mathscr{P}_{nCir} \subseteq \mathscr{P}_{nBP}$ (Lemma 3) concluding the proof of Theorem 2.2.2 .

LEMMA 2.

Each nondeterministic polynomial time-bounded nonuniform Turing machine can be simulated by a sequence of nondeterministic polynomial size circuits, i.e.

$$\mathscr{NP} \subseteq \mathscr{P}_{nCir} .$$

PROOF.

If $A \in \mathcal{NP} \ (= NP/n^{O(1)})$ then $A = \{w \mid w\#\alpha(|w|) \in A\}$ for an advice α of polynomial length and some $A' \in NP$. Describing A' by means of polynomially bounded quantifiers

$$A' = \{ u \mid \exists \, y \in \{0,1\}^{p(|u|)} : u\#y \in A'' \}$$

for a polynomial p and some $A'' \in P$ we get

$$A' = \{w\#\alpha(|w|) \mid \exists \, y \in \{0,1\}^{p(|w\#\alpha(|w|)|)} : w\#\alpha(|w|)\#y \in A''\}.$$

Since $A'' \in P$ there is a sequence of Boolean circuits $\{C_m\}$ of polynomial size in the input-length $m = n + p(n) + |\alpha(n)|$ which is polynomial in n, too. Working on the $n + p(n)$ input variables (w,y) and taking the advice $\alpha(n)$ as constant input circuit C_m computes 1 iff $w\#\alpha(|w|)\#y \in A''$. If we consider C_m, $m = n + p(n) + |\alpha(n)|$, as a nondeterministic circuit C_n' of the first n variables we have

$$\chi_{A^n}(w) \ = \ \bigvee_{y \in \{0,1\}^{p(n)}} C_n(w,y)$$

for $w \in \{0,1\}^*$, $|w| = n$. Consequently, C_n' computes A^n nondeterministically and we obtain $\mathcal{NP} \subseteq \mathcal{P}_{nBP}$. (For a more detailed description of this simulation we refer to [Sch85].) □

LEMMA 3.

Each nondeterministic circuit of polynomial size can be simulated by a polynomial size nondeterministic branching program, i.e.

$$\mathcal{P}_{nCir} \ \subseteq \ \mathcal{P}_{nBP} \ .$$

PROOF.

Let C_n be a polynomial size nondeterministic circuit computing a set A^n. A nondeterministic branching program for A^n

can be constructed in the following way. First we assign a Boolean variable y_v to each node v of C in such a way that

(i)   if v is an internal node then y_v is a nondeterministic variable, and

(ii) if v is an input node of C, then y_v is the deterministic or nondeterministic variable that input node is labelled with.

For each node v of C with predecessors $v_1,\ldots, v_r$, we can compute the Boolean equivalences

$$y_v \longleftrightarrow y_{v_1} \wedge \ldots \wedge y_{v_r} \qquad \text{if } v \text{ is an } \wedge\text{-node,}$$

or

$$y_v \longleftrightarrow y_{v_1} \vee \ldots \vee y_{v_r} \qquad \text{if } v \text{ is an } \vee\text{-node}$$

by means of the following branching programs $P_v^{\vee}$ and $P_v^{\wedge}$:

$P_v^{\wedge}$:   and $P_v^{\vee}$:

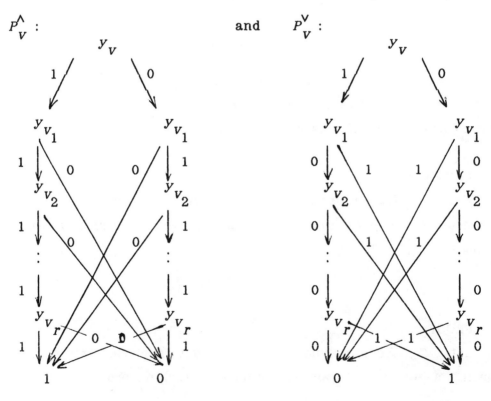

Obviously, a nondeterministic branching program which checks whether all these equivalences hold simultaneously simulates C_n . In order to construct such a nondeterministic branching program P_n let $\leq$ be a linear order on the set V of the nodes v of C_n with $v \leq v'$ if v' is a predecessor of v in C . Now P_n can be constructed from the $P_v^\wedge$ and $P_v^\vee$, $v \in V$, by identifying the source of the corresponding P_v^ω , $\omega \in \{\wedge,\vee\}$, for each $v \in V$, with the 1-sink of the corresponding $P_{v'}^{\omega'}$, $\omega' \in \{\wedge,\vee\}$, if v' is covered by v in the linear order $\leq$.

Since P_n simulates C_n and is of polynomial size $O(Size(C_n)^2)$ we are done. $\square$

This finishes the proof of Theorem 2.2.2 . ∎

2.3. NONDETERMINISTIC BOUNDED WIDTH BRANCHING PROGRAMS OF POLYNOMIAL SIZE

In this section we investigate the influence of bounding the width on the computational power of nondeterministic branching programs. Due to the classification of Section 2.1 grouping nondeterministic branching programs into 1-time-only-nondeterministic branching programs and (unrestricted) nondeterministic branching programs we establish a different behavior of these two types of nondeterministic branching programs if we bound their width. While nondeterministic bounded width branching programs turn out to be as powerful as nondeterministic branching programs without any width constraints (Paragraph 2.3.2), 1-time-only-nondeterminism does not even increase the computational power of (ordinary) bounded width branching programs (Paragraph 2.3.1).

Both results were first published in [Me87,3].

2.3.1. 1-TIME-ONLY-NONDETERMINISTIC BOUNDED WIDTH BRANCHING PROGRAMS OF POLYNOMIAL SIZE

In the following we prove that 1-time-only-nondeterminism does not increase the computational power in the case of bounded width branching programs. Due to the result of Barrington [Ba86] which states that bounded width polynomial size (ordinary) branching programs are of the same computational power as Boolean (fan-in 2) circuits of logarithmic depth we obtain

$$\mathcal{P}_{bw\ n_1BP} = \mathcal{N}\mathcal{C}^1 .$$

where $\mathscr{P}_{bw\,n_1BP}$ denotes the class of languages $A \subseteq \{0,1\}^*$ accepted by (sequences of) 1-time-only-nondeterministic polynomial size branching programs of bounded width.

THEOREM 2.3.1.

Sequences of polynomial size 1-time-only-nondeterministic branching programs of bounded width and sequences of Boolean (fan-in 2) circuits of logarithmic depth are of the same computational power. I.e.

$$\mathscr{P}_{bw\,n_1BP} = \mathscr{NC}^1 .$$

PROOF.

Due to the result of Barrington it suffices to prove $\mathscr{P}_{bw\,BP} = \mathscr{P}_{bw\,n_1BP}$. Trivially, we only have to show $\mathscr{P}_{bw\,n_1BP} \subseteq \mathscr{P}_{bw\,BP}$.

Let $A \in \mathscr{P}_{bw-n_1BP}$. Then there exists $w' \in \mathbb{N}$ such that every restriction A^n of A will be accepted by a 1-time-only-nondeterministic branching program P' of width w' and length l', polynomial in n. Since, obviously, two 1-time-only-nondeterministic branching programs which differ only in the nondeterministic variables assigned to the nodes compute the same function, we can assume that every nondeterministic variable of P' is assigned exactly once to a node of P'. Thus, by means of the construction of Section 1.3, we can transform P' into a normal form 1-time-only-nondeterministic polynomial size branching program P of width w, $w \leq 2w'$, and length l, $l \leq (n+1) \cdot l'$, which accepts A^n.

Obviously, a level j, $0 \leq j < l-1$, of P labelled by the (deterministic or nondeterministic) variable z_i is completely described by the two functions

$$f_j \ , \ g_j : \quad [w] \ \longrightarrow \ [w]$$

where $[w] := \{1,...,w\}$, which give the end points in level $j+1$ of the two edges leaving the nodes of level j : $f_j(v)$ is the end point of the edge starting in the node $v \in [w]$ of level j which corresponds to $z_i = 0$ and $g_j(v)$ is the end point of the edge corresponding to $z_i = 1$. The last but one level $l-1$ can be described by the two functions

$$f_{l-1} \ , \ g_{l-1} \quad : \quad [w] \ \longrightarrow \ \{0,1\}$$

indicating the sinks to which the nodes of level $l-1$ are connected if $z_i = 0$ and $z_i = 1$, respectively.

Now we simulate the 1–time–only–nondeterministic branching program P level by level by an ordinary (deterministic) branching program P'' of width 2^W whose length equals that of P , what implies $A \in \mathcal{P}_{bw-BP}$. In order to do this, we identify the nodes of each level of P'' with the elements M of $2^{[w]}$. If level j , $0 \le j < l$, of P is labelled by the deterministic variable x_i , then we label the nodes of level j of P'' by x_i too, and define

$$f_j \ , \ g_j : \ 2^{[w]} \ \longrightarrow \ \begin{cases} 2^{[w]} & \text{if } 0 \le j < l-1 \ ; \\ 2^{\{0,1\}} & \text{if } j = l \ . \end{cases}$$

for $M \in 2^{[w]}$ by

$$f_j(M) = \{f_j(m) \mid m \in M\} \text{ and } g_j(M) = \{g_j(m) \mid m \in M\}.$$

But if level j of P is labelled by the nondeterministic variable y_i , then we label the nodes of level j of P'' by any one of the deterministic variables $x_1,...,x_n$ and define

$$f_j(M) \;=\; g_j(M) \;=\; \bigcup_{m \in M} \{f_j(m)\} \;\cup\; \{g_j(m)\} \;.$$

Taking in P'' the node $\{0\}$ as 0-sink and the union of the nodes $\{1\}$ and $\{0,1\}$ as 1-sink we obtain a polynomial size branching program of bounded width which accepts exactly A^n .∎

2.3.2. NONDETERMINISTIC BOUNDED WIDTH BRANCHING PROGRAMS OF POLYNOMIAL SIZE

Contrary to the case of 1-time-only-nondeterministic branching programs we will prove in this section that restricting the width does not decrease the computational power of (general) nondeterministic branching programs. Thus, nondeterministic bounded width branching programs of polynomial size are as powerful as nondeterministic polynomial time-bounded nonuniform Turing machines. Moreover, in the following section we will also prove that 2-times-only-nondeterministic polynomial size branching programs of width 3 are already of the same computational power as general nondeterministic polynomial size branching programs. This result, first published in [Me87,3], sharpens a result of Valiant [Va81] which describes $\mathcal{NP}$ by means of polynomial size nondeterministic formulas.

In the following we denote the class of languages acceptable by (sequences of) nondeterministic polynomial size branching programs of width w by $\mathcal{P}_{width-w\ nBP}$. Further, $\mathcal{P}_{width-w\ n_r BP}$ denotes the class of languages acceptable by sequences of r-times-only-nondeterministic width w branching programs of polynomial size.

THEOREM 2.3.2.

Sequences of polynomial size nondeterministic branching programs of bounded width and nonuniform polynomial time–bounded nondeterministic Turing machines are of the same computational power, i.e.

$$\mathcal{P}_{bw\ nBP} = \mathcal{N}\mathcal{P} .$$

PROOF.

Due to Theorem 2.2.2 it suffices to prove $\mathcal{P}_{bw\ nBP} = \mathcal{P}_{nBP}$. Trivially, we only have to show that $\mathcal{P}_{nBP} \subseteq \mathcal{P}_{bw\ nBP}$.

In order to do this we proceed in the following way: First we prove that polynomial size nondeterministic branching programs can be simulated by polynomial size nondeterministic circuits (Lemma 4). Then we show how to simulate polynomial size nondeterministic circuits by polynomial size nondeterministic circuits of depth 2 (Lemma 5). Finally, we simulate these circuits by nondeterministic branching programs of width 3 (Lemma 6) thus concluding the proof of Theorem 2.3.2 .

LEMMA 4.

Each bounded width nondeterministic branching program can be simulated by a nondeterministic Boolean circuit of the same size, to within a constant factor. Hence,

$$\mathcal{P}_{bw\ nBP} \subseteq \mathcal{P}_{nCir}$$

PROOF.

Adapting a construction for deterministic branching programs [We87], from a nondeterministic branching program P one obtains a nondeterministic $\{sel\}$–circuit C_P computing the same set as P . The function sel is defined by

$$sel(x,y,z) \;=\; \overline{x} \wedge y \;\vee\; x \wedge z \quad \text{for all} \quad x,y,z \in \{0,1\}$$

C_P is constructed from P by reversing the directions of all edges of P, labelling each node v of P with sel and providing it with a new predecessor, namely the circuit input node of the variable z_i by which the node v is labelled with in P. The descendant of v which was reached in P if $z_i = 0$ is taken as the second predecessor and the descendant which was reached if $z_i = 1$ is taken as the third. Replacing all sel-nodes in C_P by the following 3-node subcircuit C_{sel}

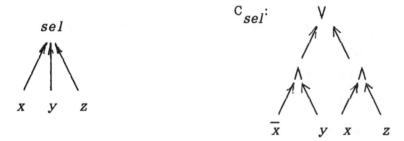

one obtains a $\{\vee,\wedge,\neg\}$-circuit C'_P with

$$Size(C'_P) \;=\; 3 \cdot Size(C_P) \;=\; 3 \cdot Size(P) \; ,$$

which computes the same set as C_P and, therefore, as P. $\square$

Let $\mathscr{F}_{depth-2,nCir}$ denote the class of sets computable by (sequences of) polynomial size nondeterministic circuits of unbounded fan-in and depth 2 .

LEMMA 5.

Each nondeterministic polynomial size circuit can be simulated by a nondeterministic unbounded fan-in circuit of polynomial size and depth 2 , i.e.

$$\mathcal{P}_{nCir} \subseteq \mathcal{P}_{depth-2,nCir} \quad .$$

PROOF.

Let C be a nondeterministic circuit computing a set A^n. From C we obtain a nondeterministic unbounded fan-in circuit C' of depth 2 which computes A^n in the following way. First we assign a Boolean variable $\alpha(v)$ to each node v of C such that

(i)   If v is an internal node, we assign to v a nondeterministic variable $y_v = \alpha(v)$, and

(ii) if v is an input node of C, then we assign to v the deterministic or nondeterministic variable with which v is labelled.

Then we compute at a node v with predecessors $v_1,..., v_r$ the Boolean equivalence

$$y_v \;\leftrightarrow\; \alpha(v_1) \wedge ... \wedge \alpha(v_r) \quad \text{if} \quad v \text{ is an } \wedge\text{-node},$$

or

$$y_v \;\leftrightarrow\; \alpha(v_1) \vee ... \vee \alpha(v_r) \quad \text{if} \quad v \text{ is an } \vee\text{-node}.$$

Together with the final check whether all equivalences hold, these computations can be done in parallel by a nondeterministic circuit of polynomial size and depth 2. $\square$

LEMMA 6.

Each nondeterministic unbounded fan-in circuit of polynomial size and depth 2 can be simulated by a nondeterministic polynomial size branching program of width 3 , i.e.

$$\mathcal{P}_{depth-2,nCir} \subseteq \mathcal{P}_{width-3,nBP} \quad .$$

PROOF.

Again we can adapt a construction of Wegener [We87] for getting width-2 branching programs from depth-2 circuits.

Obviously, the following nondeterministic, width-2 branching programs P_c and P_d

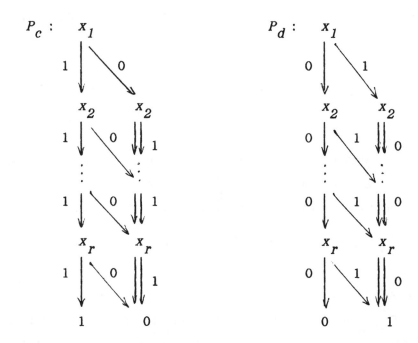

compute the conjunction $c = x_1 \wedge x_2 \wedge \ldots \wedge x_r$ and the disjunction $d = x_1 \vee x_2 \vee \ldots \vee x_r$, respectively.

Let C be a nondeterministic, depth-2 circuit whose least gate is a conjunction (similar arguments work for a disjunction). If $d_1, \ldots, d_m$ are the disjunctions computed in the first logical level of C, then we can simulate C by a nondeterministic branching program P_C constructed from the nondeterministic branching programs $P_{d_1}, \ldots, P_{d_m}$ by identifying the 1-sink of P_{d_i} with the source of $P_{d_{i+1}}$ for each i, $1 \le i < m$. Since P_C is of width 3 and of polynomial size in the number of gates of C after a transformation in normal form we are done. $\square$

This finishes the proof of Theorem 2.3.2 . ∎

Let us only remark that, if the branching programs under consideration are not required to be levelled then the construction of Lemma 6 provides a nondeterministic branching program of width merely 2 .

Indeed, we can prove even more.

COROLLARY 7.

Each polynomial size nondeterministic branching program of width w can be simulated by a polynomial size 2-times-only-nondeterministic branching program of width w , i.e.

$$\mathcal{P}_{width-w,nBP} \quad = \quad \mathcal{P}_{width-w,n_2BP} \quad , \ w \geq 2 \ . \quad \blacksquare$$

PROOF.

Let P be a nondeterministic width-w branching program. We first obtain a 1-time-only-nondeterministic branching program P' of width w by replacing the variables z_i assigned to nodes v of P by new nondeterministic variables $z_{i,v}$. We then get a 2-times-only-nondeterministic branching program P'' accepting the same set as P , if we also check all Boolean equivalences

$$(\bigwedge_{v \in P} z_{i,v}) \longleftrightarrow z_i$$

for all nondeterministic variables z_i before accepting an input. Since these computations can be done by polynomial-size depth-2 circuits we can, due to Lemma 5 check these equivalences by polynomial-size branching programs of width 2 . □

Corollary 7 along with Theorem 2.3.2 yields

COROLLARY 8.

Sequences of polynomial size 2−times−only−nondeterministic branching programs of width 3 and nondeterministic polynomial time−bounded nonuniform Turing machines are of the same computational power, i.e.

$$\mathcal{P}_{width-3,n_2BP} = \mathcal{NP} \cdot \blacksquare$$

Altogether we have proved that 2−times−only−nondeterministic branching programs of width 3 are as powerful as unrestricted nondeterministic branching programs of polynomial size. It should be mentioned that we could further sharpen this result replacing 2−times−only−nondeterministic polynomial size branching programs of width 3 by 2−times−only−nondeterministic, polynomial size branching programs of width 3 which are 1−time−only in all deterministic variables.

Chapter 3

Ω – BRANCHING PROGRAMS AND THEIR COMPUTATIONAL POWER

Introduction

Ω-branching programs, introduced in [Me87,1], generalize the concept of branching programs by equipping some of the nodes with devices for evaluating Boolean functions ω, $\omega \in \Omega$, from a set $\Omega \subseteq \mathbb{B}_2$ of 2-argument Boolean functions. E.g. $\emptyset$-branching programs are ordinary branching programs while $\{\vee\}$-branching programs are computationally and structurally equivalent to 1-time-only-nondeterministic branching programs (Proposition 1). Having in mind the results of Chapter 2 already these few examples would suggest the conjecture that Ω-branching programs work more efficiently for suitable $\Omega \subseteq \mathbb{B}_2$ than ordinary branching programs. The purpose of this chapter is to investigate this question.

After classifying in Section 3.1 Ω-branching programs, $\Omega \subseteq \mathbb{B}_2$, into the five types of (ordinary) branching programs, disjunctive $\{\vee\}$-branching programs, conjunctive $\{\wedge\}$-branching programs, parity $\{\oplus\}$-branching programs and alternating $\{\vee,\wedge\}$-branching programs (Theorem 3.1) [Me88] we study the behavior of these types of Ω-branching programs under different resource constraints.

At first, in Section 3.2, we let the size of our Ω-branching programs be polynomially bounded. Introducing the concept of Ω-Turing machines, $\Omega \subseteq \mathbb{B}_2$, [Me88], which generalizes that of alternating Turing machines we can relate the class of languages acceptable by (sequences of) polynomial size Ω-branching programs, $\Omega \subseteq \mathbf{B}_2$, to the class of languages computable by logarithmic space-bounded nonuniform Ω-Turing machines (Theorem 3.2.1). If $\mathcal{P}_{\Omega-BP}$ denotes this class, then Theorem 3.2.2 will imply the relations

$$\mathcal{P}_{\{\vee\}-BP} = \mathcal{NL} ,$$

$$\mathcal{P}_{\{\wedge\}-BP} = co\text{-}\mathcal{NL} (= \mathcal{NL} \text{ [Im87, Sz87]}) , \text{ and}$$

$$\mathcal{P}_{\{\vee,\wedge\}-BP} = \mathcal{P} ,$$

in addition to the classical result $\mathcal{P}_{BP} = \mathcal{L}$ of Corollary 1, Chapter 1).

However, the remaining fifth class $\mathcal{P}_{\{\oplus\}-BP}$, according to the classification result of Theorem 3.1 , has not been identified up to now in the context of logarithmic space-bounded Turing machines although it seems to be as interesting as the other ones.

In Section 3.3 we study polynomial size Ω-branching programs of bounded width. We prove that equipping the nodes of a bounded width branching program with devices for evaluating Boolean functions from a set $\Omega \subseteq \mathbf{B}_2$ does not increase their computational power. Generalizing the result of Barrington [Ba86] we find that each polynomial size and bounded width Ω-branching program, $\Omega \subseteq \mathbf{B}_2$, may be simulated by a Boolean (fan-in 2) circuit of logarithmic depth (Theorem 3.3). Let us just mention in this context that width-restricted alternating branching programs seem to be a more natural model than the computationally equivalent (Corollary 11) width restricted Boolean circuits of Hoover and Barrington [Ba86].

Section 3.4 is devoted to the study of quasipolynomial size Ω-branching programs, $\Omega \subseteq \mathbf{B}_2$. Quasipolynomial size Ω-branching programs are Ω-branching programs of size $2^{(logn)^{O(1)}}$. Interest in these devices and in the classes $\mathcal{Q}_{\Omega-BP}$ of languages acceptable by sequences of such quasipolynomial size Ω-branching programs arises from the fact that they are very resistant to changes in the set $\Omega \subseteq \mathbb{B}_2$ and to bounding their width. So, we can prove in Theorem 3.4.2 that, whenever $\Omega \cup \{0, 1, id_l, id_r\}$ is not complete and $\Omega \neq \{\vee, \wedge\}$, all the classes $\mathcal{Q}_{\Omega-BP}$, $\Omega \subseteq \mathbb{B}_2$, coincide. Moreover, these classes coincide with the class $\Omega_{bw\ \Omega-BP}$ of languages acceptable by sequences of bounded width quasipolynomial size Ω-branching programs, a result which is very unlikely to be true in the corresponding case of polynomial size Ω-branching programs.

One of the most important problems in complexity theory is that of separating complexity classes such as $\mathcal{L}$, $\mathcal{NL}$ or $\mathcal{P}$ (or to prove their coincidence). In Section 3.2 we have given a description of these complexity classes by means of certain polynomial size Ω-branching programs. Hence superpolynomial lower bounds for branching programs, disjunctive $\{\vee\}$-branching programs and conjunctive $\{\wedge\}$-branching programs would essentially contribute to a separation of these classes. However, up to now exponential lower bounds for branching programs could be obtained only for read-once-only branching programs (or similar for real time branching programs) [We84], [Ža84], [A&86], [KW86], [Kr87]. In order to separate larger classes we investigate read-once-only Ω-branching programs, $\Omega \subseteq \mathbb{B}_2$, in the final Section 3.5 . This approach indeed has proved to be quite successful since it enables us to separate the complexity classes $\mathcal{P}_{\Omega-BP1}$, $\Omega \subseteq \mathbb{B}_2$, related to polynomial size read-once-only Ω-branching programs (Theorem 3.5.4). Most of these separation results are proved by means of an exponential lower bound for the problem of deciding whether a given Boolean matrix is a permutation matrix (Lemma 31). This lower bound was obtained in cooperation with M.Krause and S.Waack [KMW88].

Since the read-once-only Ω-branching program complexity classes $\mathcal{P}_{BP1}$, $\mathcal{P}_{\{\vee\}-BP1}$, $\mathcal{P}_{\{\wedge\}-BP1}$ and $\mathcal{P}_{\{\vee,\wedge\}-BP1}$ correspond to the known nonuniform logarithmic space-bounded eraser Turing machine classes $\mathcal{L}_e$, $\mathcal{NL}_e$, $co\text{-}\mathcal{NL}_e$ and $\mathcal{P}_e$ (Theorem 3.5.1) we obtain

$$
\begin{array}{ccccc}
 & & \mathcal{NL}_e & & \\
 & \underset{\neq}{\subset} & & \underset{\neq}{\subset} & \\
\mathcal{L}_e & & & & \mathcal{P}_e = \mathcal{P} . \\
 & \underset{\neq}{\subset} & & \underset{\neq}{\subset} & \\
 & & co\text{-}\mathcal{NL}_e & &
\end{array}
$$

Since up to now only $\mathcal{L}_e = \mathcal{NL}_e$ has been separated by exponential lower bounds with $\mathcal{L}_e \underset{\neq}{\overset{\subset}{}} \mathcal{NL}_e$, $\mathcal{L}_e \underset{\neq}{\overset{\subset}{}} co\text{-}\mathcal{NL}_e$ and $\mathcal{NL}_e \underset{\neq}{\overset{\subset}{}} \mathcal{P}_e$, $co\text{-}\mathcal{NL}_e \underset{\neq}{\overset{\subset}{}} \mathcal{P}_e$ we have taken further steps in separating larger and larger complexity classes by means of exponential lower bounds. On the other hand, we obtain

$$\mathcal{NL}_e \underset{\neq}{\overset{\subset}{}} \mathcal{NL} \quad \text{and} \quad co\text{-}\mathcal{NL}_e \underset{\neq}{\overset{\subset}{}} co\text{-}\mathcal{NL} = \mathcal{NL}$$

as a corollary of $\mathcal{NL}_e \neq co\text{-}\mathcal{NL}_e$ and of the Immerman/ Szelepcsenyi result $\mathcal{NL} = co\text{-}\mathcal{NL}$ [Im87, Sz87]. This proves that the eraser concept causes proper restrictions of the computational power not only in the deterministic case but also in the nondeterministic and in the co-nondeterministic cases. Similarly we obtain

$$\mathcal{P}_{\{\vee\}-BP1} \underset{\neq}{\overset{\subset}{}} \mathcal{P}_{\{\vee\}-BP} \quad \text{and} \quad \mathcal{P}_{\{\wedge\}-BP1} \underset{\neq}{\overset{\subset}{}} \mathcal{P}_{\{\wedge\}-BP} = \mathcal{P}_{\{\vee\}-BP}$$

which proves that read-once-only disjunctive and read-once-only conjunctive branching programs are less powerful than those not assumed to be read-once-only.

3.1. Ω-BRANCHING PROGRAMS AND THEIR CLASSIFICATION

The purpose of this section is to start the study of Ω-bran-ching programs, $\Omega \subseteq \mathbb{B}_2$. In Paragraph 3.1.1 we introduce Ω-branching programs and prove the computationally equivalence of disjunctive $\{v\}$-branching programs and 1-time-only-nonde-terministic branching programs investigated in Chapter 2. Then, in Section 3.1.2 , we give a complete classification of Ω-branching programs (Theorem 3.1.2). We prove that each Ω-branching program is computationally equivalent either to an (ordinary) branching program, to a disjunctive $\{v\}$-branching program, to a conjunctive $\{\wedge\}$-branching program, to a parity $\{\oplus\}$-branching program or to an alternating $\{v,\wedge\}$-branching program.

3.1.1. Ω – BRANCHING PROGRAMS

An Ω-*branching program* P is a branching program some of whose non-sink nodes are equipped with devices for evaluating Boolean functions $\omega \in \Omega$ from a set $\Omega \subseteq \mathbb{B}_2$ of 2-argument Boolean functions. Formally, this can be described by labelling some of the non-sink nodes of P by Boolean functions $\omega \in \Omega$ instead of Boolean variables. The Boolean values assigned to the sinks of P extend to Boolean values associated with all nodes of P in the following way: if the both successor nodes v_0 and v_1 of a node v of P carry the Boolean values δ_0 and δ_1 , respectively, and if v is labelled by a Boolean variable x_i we associate with v the value δ_0 or δ_1 iff $x_i = 0$ or $x_i = 1$. If v is labelled by a Boolean function

ω then we associate with v the value $\omega(\delta_0, \delta_1)$. P is said to *accept* (*reject*) an input $w \in \{0,1\}^n$ if the source of P is associated with 1 (0) under w .

An Ω-branching program P and an Ω'-branching program P' , Ω, $\Omega' \subseteq \mathbb{B}_2$, are said to be *computationally equivalent* if they accept the same set and if their sizes coincide, to within a constant factor.

Ω-branching programs with $\Omega = \{\vee\}$, $\{\wedge\}$, $\{\oplus\}$ and $\{\vee, \wedge\}$ are called *disjunctive*, *conjunctive*, *parity* and *alternating* branching programs.

Due to the following proposition Ω-branching programs are generalizations of 1-time-only-nondeterministic branching programs. These two concepts coincide in the case of $\Omega = \{\vee\}$.

PROPOSITION 1.

For each disjunctive $\{\vee\}$-*branching program there is a computationally equivalent* 1-*time-only-nondeterministic branching programs and vice versa.*

PROOF.

Let P be a disjunctive $\{\vee\}$-branching program. Replacing each $\vee$-node v of P by a nondeterministic variable y_v we obtain a nondeterministic branching program P' of equal size. Obviously, P' accepts the same set as P and is 1-time-only-nondeterministic.

Reversely, if we replace all nondeterministic variables of a given 1-time-only-nondeterministic branching program P by the Boolean function $\vee$ then we obtain a disjunctive $\{\vee\}$-branching program P' of the same size. If P accepts an input w then, by definition of nondeterministic branching programs, there is a setting of the nondeterministic variables such that the path traced under w is an accepting one. This and the property of P to be 1-time-only-nondeterministic imply that the root of P' is associated with 1 under w. Hence P and P' accept the same set. ∎

3.1.2. CLASSIFICATION

In the following we completely classify all Ω-branching programs.

At first we observe that the Boolean functions 0, 1, id_l, id_r (see the table in Figure 3.1) belong to the basic equipment of every Ω-branching program.

PROPOSITION 2.

Let Ω_0 be the set $\Omega_0 = \{0, 1, id_l, id_r\} \subseteq \mathbf{B}_2$. Each $(\Omega \cup \Omega_0)$-branching program P may be simulated by an Ω-branching program P' of at most equal size.

PROOF.

Let P be an $(\Omega \cup \Omega_0)$-branching program. We easily obtain an Ω-branching program P' which accepts the same set as P if we replace all nodes labelled by 0 (1) by 0-sinks $(1$-sinks$)$, and if we identify a node v with its left (right) successor whenever this is labelled by id_l (id_r). Obviously,

$$Size(P') \leq Size(P) . \quad \blacksquare$$

Now let us consider complete bases in $\mathbf{B}_2$. A set $\Omega \subseteq \mathbf{B}_2$ of Boolean functions is called a *complete basis* if any Boolean function can be computed by means of the elements of Ω.

PROPOSITION 3.

Let $\Omega \subseteq \mathbb{B}_2$ *be a complete basis . For each* Ω-*branching program there is a computationally equivalent alternating* $\{\vee,\wedge\}$-*branching program.*

PROOF.

Let $\Omega \subseteq \mathbb{B}_2$ be a complete basis and let P be an Ω-branching program. Further let $sel = sel(x,y,z)$ be the Boolean function defined by

$$sel(x,y,z) = (\overline{x} \wedge y) \vee (x \wedge z) \quad \text{for} \quad x, y, z \in \{0,1\} .$$

Adapting a construction for ordinary branching programs [We87] from P we obtain an $(\Omega \cup \{sel\})$-circuit C_P which accepts the same set as P. In detail, C_P is constructed from P by reversing the directions of all edges of P and labelling the ω-nodes, $\omega \in \Omega$, of P by ω. The remaining nodes v are labelled by sel and get a new predecessor, namely the circuit input node of the variable x_i by which v is labelled in P. The descendant of v which is reached in P if $x_i = 0$ is taken as the second predecessor and the descendant which is reached in P if $x_i = 1$ is taken as the third. Obviously, C_P computes the same set as P and its size equals that of P.

However, C_P can be simulated by an $\{\wedge,\vee\}$-circuit C_P' of size

$$Size(C_P') = k \cdot Size(C_P)$$

for a constant $k \in \mathbb{N}$ on the basis of a well-known standard argument which can be found for instance in [Sa76].

Finally, by means of the following construction we obtain $\{\wedge,\vee\}$-branching program P' from C_P' which simulates C_P': Reverse the directions of all edges of C_P' and replace the input nodes x_i and $\overline{x}_i$, $1 \le i \le n$, by the 1-node branching

programs

and

respectively.

Altogether, P' is an alternating $\{\wedge,\vee\}$-branching program of size

$$Size(P') = Size(C_P')$$
$$\leq k \cdot Size(C_P)$$
$$= k \cdot Size(P) .$$

which simulates P.

Reversely, if $\Omega \subseteq \mathbf{B}_2$ is a complete basis we can compute $\vee$ and $\wedge$ from functions of Ω. Replacing the disjunctive $\vee$-nodes and the conjunctive $\wedge$-nodes by Ω-subcircuits which perform these computations we obtain, from a given $\{\vee,\wedge\}$-branching program P, an Ω-branching program P' of size

$$Size(P') \leq k \cdot Size(P) , \quad k \in \mathbf{N} ,$$

which simulates P. ∎

Due to Proposition 3 , for $\Omega \subseteq \mathbf{B}_2$ and $\Omega \cup \Omega_0$ complete, we can restrict investigations of Ω-branching programs to alternating $\{\wedge,\vee\}$-branching programs.

What happens if $\Omega \cup \Omega_0 \subseteq \mathbf{B}_2$ is not complete?

If $\mathbf{M}$, $\mathbf{L}$, $\mathbf{S}$, $\mathbf{T}_0$ and $\mathbf{T}_1$ denote the classes of *monotone*, *linear*, *self-dual*, 0-*preserving* and 1-*preserving* functions of $\mathbf{B}_2$, then a classical Theorem of Post [Po21] states that $\Omega \cup \Omega_0$ must be contained in one of the above-mentioned five classes:

$$\Omega \cup \Omega_0 \subseteq \mathbf{M} \ , \ \Omega \cup \Omega_0 \subseteq \mathbf{L} \ , \ \Omega \cup \Omega_0 \subseteq \mathbf{S} \quad \text{or} \quad \Omega \cup \Omega_0 \subseteq \mathbf{T}_i \ (i=0,1).$$

The table of Figure 3.1 shows:

$$\mathbf{M} = \Omega_0 \cup \{ \vee, \wedge \} \ ,$$

$$\mathbf{L} = \Omega_0 \cup \{ \ \neg_l, \ \neg_r, \ \leftrightarrow, \ \oplus \} \ ,$$

$$\mathbf{S} = \{ id_l, \ id_r, \ \neg_l, \ \neg_r \} \ ,$$

$$\mathbf{T}_0 = \{ \ 0, \ id_l, \ id_r, \ \oplus, \ \vee, \ \wedge, \ \not\Rightarrow, \ \not\Leftarrow \} \quad \text{and}$$

$$\mathbf{T}_1 = \{ \ 1, \ id_l, \ \ id_r, \ \leftrightarrow, \ \vee, \ \wedge, \ \Rightarrow, \ \Leftarrow \}.$$

	Function	monotone	linear	self-dual	T_0	T_1
0	false	1	1	0	1	0
1	true	1	1	0	0	1
id_l	left identity	1	1	1	1	1
id_r	right identity	1	1	1	1	1
$\neg_l$	left negation	0	1	1	0	0
$\neg_r$	right negation	0	1	1	0	0
$\leftrightarrow$	equivalence	0	1	0	0	1
$\oplus$	exclusive or	0	1	0	1	0
$\wedge$	and	1	0	0	1	1
$\vee$	or	1	0	0	1	1
$\uparrow$	nand	0	0	0	0	0
$\downarrow$	nor	0	0	0	0	0
$\Rightarrow$	implies	0	0	0	0	1
$\not\Rightarrow$	not implied	0	0	0	1	0
$\Leftarrow$	implied by	0	0	0	0	1
$\not\Leftarrow$	not implied by	0	0	0	1	0

Figure 3.1. Table of the set $\mathbf{B}_2$ of all 2-argument Boolean functions.

Since $\Omega \cup \Omega_0 \nsubseteq S$, T_0 , T_1 we get

$$\Omega \subseteq M \quad \text{or} \quad \Omega \subseteq L \quad .$$

First let $\Omega \subseteq M = \Omega_0 \cup \{\vee, \wedge\}$. Then, due to Proposition 2 it suffices to consider the following four types of Ω-branching programs:

- (ordinary) branching programs ($\Omega = \emptyset$),

- disjunctive $\{\vee\}$-branching programs,

- conjunctive $\{\wedge\}$-branching programs, and

- alternating $\{\vee,\wedge\}$-branching programs.

Now let $\Omega \subseteq L = \{\neg_l, \neg_r, \leftrightarrow, \oplus\}$. Obviously, we can think of the 2-input $\neg_l$-nodes and $\neg_r$-nodes as of 1-input $\neg$-nodes.

PROPOSITION 4.

Let $\Omega \subseteq L = \{\neg_l, \neg_r, \leftrightarrow, \oplus\}$. Each Ω-branching program may be simulated both by a parity $\{\oplus\}$-branching program and by a $\{\leftrightarrow\}$-branching program of the same size, to within a constant factor.

PROOF.

Using the identities

$$\neg a = a \oplus 1 \quad \text{and} \quad (a \leftrightarrow b) = a \oplus (b \oplus 1)$$

we can convert an Ω-branching program P , $\Omega \subseteq L$, into a parity $\{\oplus\}$-branching program P' of size

$$Size(P') \leq 2 \cdot Size(P) \quad .$$

Analogously, we may obtain a $\{\leftrightarrow\}$-branching program by means of the identities

$$\neg a = (a \leftrightarrow 0) \quad \text{and} \quad a \oplus b = a \leftrightarrow (b \leftrightarrow 0) . \quad \blacksquare$$

Due to the Propositions 2 and 4 each Ω-branching program, $\Omega \subseteq \mathbf{L}$, is computationally equivalent to

- an (ordinary) branching program,

- a $\{\neg\}$-branching program, or

- a parity $\{\oplus\}$-branching program.

However, $\{\neg\}$-branching programs are no more powerful than ordinary ones:

PROPOSITION 5.

For each $\{\neg\}$-branching program there is a computationally equivalent (ordinary) branching program.

PROOF.

In order to construct an ordinary branching program $\overline{P}$ which simulates P we take two copies P' and P'' of P in order to remember the parity of the number of negations that have been passed. If v is a node of P then we denote the copy of v in P' by v' and the copy of v in P'' by v''. In P'' we replace the 0-sink and the 1-sink by a 1-sink and a 0-sink, respectively. Then, if v is a $\neg$-node in P with the successor node u , we delete v' in P' and v'' in P'' and connect all (direct) predecessors of v' with u'' and all (direct) predecessors of v'' with u' . Omitting all nodes not reachable from the source v_0' of P' we obtain an (ordinary) branching program $\overline{P}$ of at most doubled size. A straightforward inductive argument finally proves that $\overline{P}$ and P accept the same set. ■

Altogether we have proved the following classification theorem:

THEOREM 3.1.

For each Ω-branching program, $\Omega \subseteq \mathbb{B}_2$, there is a computationally equivalent Ω'-branching program with

$$\Omega' = \emptyset \ , \ \Omega' = \{\vee\} \ , \ \Omega' = \{\wedge\} \ , \ \Omega' = \{\oplus\} \ , \ \text{or} \ \ \Omega' = \{\vee,\wedge\} \ . \ \blacksquare$$

Hence each Ω-branching program, $\Omega \subseteq \mathbb{B}_2$, can be computationally equivalently represented either by an ordinary branching program, by a disjunctive $\{\vee\}$-branching program, a conjunctive $\{\wedge\}$-branching program, or a parity $\{\oplus\}$-branching program, or by an alternating $\{\vee,\wedge\}$-branching program.

3.2. Ω-BRANCHING PROGRAMS OF POLYNOMIAL SIZE

In the present section we examine the relationships among complexity classes defined by polynomial size Ω-branching programs. In order to do this in Paragraph 3.2.1 we introduce the concept of Ω-Turing machines , $\Omega \subseteq \mathbb{B}_2$. Briefly speaking, Ω-Turing machines are generalized alternating Turing machines which are obtained from nondeterministic Turing machines with terminating computation paths by labelling the states with Boolean functions ω of Ω instead of $\{\vee,\wedge\}$ as in the case of alternating Turing machines. Theorem 3.2.1 states that the class of languages nonuniformly computable by logarithmic space-bounded Ω-Turing machines coincides with the class of languages acceptable by sequences of polynomial size Ω-branching programs, $\Omega \subseteq \mathbb{B}_2$. If $\mathscr{P}_{\Omega-BP}$ denotes this class then, Theorem 3.2.2 of Paragraph 3.2.2 implies the relations

$$\mathscr{P}_{\{\vee\}-BP} = \mathscr{NL} \ , \quad \mathscr{P}_{\{\wedge\}-BP} = co\text{-}\mathscr{NL} = \mathscr{NL} \quad \text{and} \quad \mathscr{P}_{\{\vee,\wedge\}-BP} = \mathscr{P}$$

in addition to the classical result $\mathcal{P}_{BP} = \mathcal{L}$ (Corollary 1 of Chapter 1).

However, the remaining fifth class $\mathcal{P}_{\{\oplus\}-BP}$, according to the classification result of Theorem 3.1 , has not been identified up to now in the context of logarithmic space-bounded Turing machines although it seems to be as interesting as the other ones.

3.2.1. COMPLEXITY CLASSES RELATED TO POLYNOMIAL SIZE Ω-BRANCHING PROGRAMS

Starting from the correspondence between sequences of polynomial size branching programs and logarithmic space-bounded nonuniform Turing machines (Corollary 1 of Chapter 1) we can relate polynomial size Ω-branching programs to certain types of Turing machines. First, in [Me86,1] we have related polynomial size disjunctive $\{v\}$-branching programs (properly the computationally equivalent 1-time-only-nondeterministic branching programs) and nondeterministic logarithmic space-bounded nonuniform Turing machines. Generalizing the concept of alternating Turing machines [CKS81] to Ω-Turing machines [Me87,1] we can relate polynomial size Ω-branching programs and logarithmic space-bounded nonuniform Ω-Turing machines. Our concept of Ω-Turing machines is rather similar to that of *extended Turing machines* proposed by Goldschlager and Parberry [GP86]. However, these concepts were independently introduced and differ in the point that our Ω-Turing machines are assumed to terminate on every computation path.

Let M be a nondeterministic 1-tape Turing machine with a read-only input tape all of whose computation paths are assumed to be finite. Further, let us assume that, in every step, M has at most two nondeterministic choices. M is called an Ω-*Turing machine*, $\Omega \subseteq \mathbb{B}_2$, if the nonterminal states are labelled by Boolean functions chosen from the set $\Omega \cup \{id\}$ and if the terminal states are labelled by Boolean constants. The *configurations* of M are five-tuples containing M's current state, its label from $\Omega \cup \{id\} \cup \{0,1\}$, the working tape content and the positions of the input and working heads. A configuration C' is a *direct successor* of a configuration C on an input $w \in \{0,1\}^*$ if C' is reachable from C in one step by means of the next-move relation of M under w . For each value of the input bit read in C the number of direct successors of C equals the arity of the Boolean function assigned to the state of C . Let C_0 be the initial configuration. *Terminal configurations* are those whose states are terminal and therefore labelled by Boolean constants.

A *computation* of an Ω-Turing machine M on an input $w \in \{0,1\}^*$ can be described by means of the *computation tree* $T(M,w)$ of M on w . Its root is the tuple (C_0,w). The nodes of $T(M,w)$ are the tuples (C,w) such that the sons of (C,w) are exactly the tuples (C',w) where C' is a direct successor of C . The leaves of $T(M,w)$ are the terminal configurations. Since we generally assume that, for each input $w \in \{0,1\}^*$, all computation paths of an Ω-Turing machine terminate the computation trees $T(M,w)$ are finite.

By means of the Boolean functions $\omega \in \Omega \cup \{id\}$ included in the configurations C of the nodes (C,w) of $T(M,w)$ the Boolean values included in the leaves of $T(M,w)$ extend to Boolean values associated with all nodes of $T(M,w)$. If the root of $T(M,w)$ gets the value 1 or 0 then $T(M,w)$ is called *accepting* or *rejecting*, respectively. M *accepts* or *rejects* $w \in \{0,1\}^*$ if its computation tree $T(M,w)$ is accepting or rejecting, respectively,

$$L(M) = \{ w \mid M \text{ accepts } w \} .$$

Note, that Ω-Turing machines with

- $\Omega = \{\vee,\wedge\}$ are alternating Turing machines,

- $\Omega = \{\vee\}$ are nondeterministic Turing machines,

- $\Omega = \{\wedge\}$ are co-nondeterministic Turing machines,

and

- $\Omega = \emptyset$ are deterministic Turing machines

with terminating computation paths. For $\Omega = \{\oplus\}$ we will speak of *parity Turing machines.*

An Ω-Turing machine M is called $s(n)$ *space-bounded* if for all $w \in \{0,1\}^*$ the configurations included in the nodes of $T(M,w)$ occupy at most space $s(|w|)$. It is an immediate consequence of the finiteness of the computation paths of Ω-Turing machines that the computation trees $T(M,w)$, $w \in \{0,1\}^*$, of a $s(n)$ space-bounded Ω-Turing machine M are of depth at most $2^{O(s(n))}$. Classes $L(\Omega)$ consisting of languages acceptable by logarithmic space-bounded Ω-Turing machines are of special interest in our further considerations.

PROPOSITION 6.

(i) $L(\emptyset)$ $=$ L ,

(ii) $L(\{\vee\})$ $=$ NL ,

(iii) $L(\{\wedge\})$ $=$ $co\text{-}NL$, *and*

(iv) $L(\{\vee,\wedge\})$ $=$ $AL = P$.

PROOF.
The proof is a consequence of the following two facts:

- ∅-Turing machines, {∨}-Turing machines, {∧}-Turing machines, {∨,∧}-Turing machines are deterministic, nondeterministic, co-nondeterministic, or alternating Turing machines, with terminating computation paths, and
- every logarithmic space-bounded deterministic, nondeterministic, co-nondeterministic or alternating Turing machine can be simulated by a logarithmic space-bounded deterministic, nondeterministic, co-deterministic or alternating Turing machine, with terminating computation paths.

The last fact was proved in [CKS81] for alternating machines. Similar considerations show that it will also be true in the case of deterministic, nondeterministic and co-nondeterministic logarithmic space-bounded Turing machines . ∎

In order to relate Ω-Turing machine classes and Ω-branching programs, $\Omega \subseteq \mathbb{B}_2$, let us consider the nonuniform counterparts

$$\mathcal{L}(\Omega) \;=\; L(\Omega) \,/\, n^{O(1)}$$

of the classes $L(\Omega)$. In the nonuniform settings Proposition 6 can be restated as

COROLLARY 7.

(i) $\mathcal{L}(\emptyset) \;=\; \mathcal{L}$,

(ii) $\mathcal{L}(\{\vee\}) \;=\; \mathcal{NL}$,

(iii) $\mathcal{L}(\{\wedge\}) \;=\; co\text{-}\mathcal{NL}$, and

(iv) $\mathcal{L}(\{\vee,\wedge\}) \;=\; \mathcal{P}$. ∎

THEOREM 3.2.1.
Polynomial size Ω-branching programs and logarithmic space-

bounded nonuniform Ω*-Turing machines,* $\Omega \subseteq \mathbb{B}_2$ *, are of the same computational power. I.e.*

$$\mathcal{P}_{\Omega-BP} \quad = \quad \mathcal{L}(\Omega) \,, \quad \Omega \subseteq \mathbb{B}_2 \,.$$

PROOF.

We will only sketch the proof. For details reference is made to the proof of Theorem 2.2.1 of Chapter 2 which treats the paradigmatic case $\Omega = \{v\}$.

In order to prove $\mathcal{P}_{\Omega-BP} \subseteq \mathcal{L}(\Omega)$ $(= L(\Omega)/n^{O(1)})$ we encode a polynomial size Ω-branching program P_n which computes the restriction $A^n = A \cap \{0,1\}^n$ of a given language $A \in \mathcal{P}_{\Omega-BP}$, and take this encoding as the advice for inputs of length n . In detail, we encode each node of P_n by its number ($O(log|P_n|)$ bits), the numbers of its direct successor nodes ($O(log|P_n|)$ bits) and its label (the index of the variable tested in v or the function, it is labelled by) ($O((log\ n))$ bits). Thus, the encoding of P_n has length $O(|P_n|(log|P_n| + log\ n))$. Now it is not difficult to construct an Ω-Turing machine M which simulates P_n step by step: after storing on the working tape the encoding of the node v which has been reached M will look for the label of v at the input tape. If v is labelled by a Boolean variable then M looks at the input tape for the input bit tested at v and copies the encoding of the appropriate successor node. If v is labelled by a Boolean function $\omega \in \Omega$ then M enters an ω-state and branches, treating all direct successor nodes of v . Obviously, M nonuniformly accepts A^n and all computations of M terminate. Since all working tape inscriptions are of length $O(log|P_n|) = O(log\ n)$ we are done.

In order to prove $\mathcal{L}(\Omega) \subseteq \mathcal{P}_{\Omega-BP}$ let M be an Ω-Turing machine which nonuniformly accepts $A \subseteq \{0,1\}^*$ in space $O(log\ n)$ by means of the polynomial length-restricted advice

$\alpha(n)$. We simulate M on inputs of length n, $n \in \mathbb{N}$, by an Ω-branching program P_n in the following manner: each configuration of M is simulated by three nodes. The initial configuration yields the source and the terminal configurations labelled by 0 and 1 yield the 0-sink and the 1-sink. Let C be an ω-configuration of M, $\omega \in \Omega$, which reads the i-th input bit of the input tape. We simulate C by means of three nodes v_C, v_C^0 and v_C^1. v_C is labelled by the i-th input variable and has two successor nodes v_C^0 and v_C^1 for $x_i = 0$ and $x_i = 1$. The nodes v_C^0 and v_C^1 are ω-nodes whose successors are the nodes v_D, v_D' and v_E, v_E', respectively, which belong to the 3-nodes components simulating the direct successor configurations D, D' and E, E' of C for $x_i = 0$ and $x_i = 1$. Since the advice $\alpha(n)$ is independent from the $w \in \{0,1\}^n$ we can avoid labels x_i, $i > n$, of nodes of P_n by 'hardwiring' the advice, i.e. by identifying v_C with v_C^0 or v_C^1 in dependence on the value of x_i, $i > n$, tested in v_C.

It can easily be seen that P_n is acyclic since all computations of an Ω-Turing machine terminate. Furthermore, P_n is an Ω-branching program accepting $w \in \{0,1\}^n$ iff M accepts $w\#\alpha(n)$. Since the size of P_n does not exceed the polynomially bounded number $2^{O(\log n)}$ of configurations of M on inputs of length n we obtain $\mathcal{L}(\Omega) \subseteq \mathcal{P}_{\Omega-BP}$. ∎

3.2.2. RELATIONSHIP BETWEEN THESE COMPLEXITY CLASSES

Since the sizes of computationally equivalent Ω-branching programs coincide, to within constant factors, due to the classification result of Theorem 3.1 each polynomial size Ω-branching program is computationally equivalent to an ordinary, a disjunctive, a conjunctive, a parity or an alternating branching program of polynomial size.

Hence, we have at most five complexity classes related to polynomial size Ω-branching programs :

$$\mathscr{P}_{BP} \text{ , } \mathscr{P}_{\{\vee\}-BP} \text{ , } \mathscr{P}_{\{\wedge\}-BP} \text{ , } \mathscr{P}_{\{\oplus\}-BP} \text{ and } \mathscr{P}_{\{\vee,\wedge\}-BP} \text{ .}$$

Due to the correspondence between polynomial size Ω-branching programs and logarithmic space-bounded nonuniform Ω-Turing machines, $\Omega \subseteq \mathbf{B}_2$, proved in Paragraph 3.2.1, we also obtain a classification of Ω-Turing machine classes:

COROLLARY 8.

Each logarithmic space-bounded nonuniform Ω-Turing machine, $\Omega \subseteq \mathbf{B}_2$, is computationally equivalent to logarithmic space-bounded ordinary, nondeterministic, co-nondeterministic, parity or alternating nonuniform Turing machine. I.e there are five classes of languages $A \subseteq \{0,1\}^*$ acceptable by logarithmic space-bounded Ω-Turing machines, $\Omega \subseteq \mathbf{B}_2$:

$$\mathscr{L} \text{ , } \mathscr{L}(\{\vee\}) \text{ , } \mathscr{L}(\{\wedge\}) \text{ , } \mathscr{L}(\{\oplus\}) \text{ and } \mathscr{L}(\{\vee,\wedge\}) \text{ . } \blacksquare$$

Due to Theorem 3.2.1 and to Corollary 7 four of the five Ω-branching program complexity classes are related to well-known nonuniform Turing machine complexity classes:

COROLLARY 9.

(i) $\quad \mathscr{P}_{BP} \quad = \quad \mathscr{L}$,

(ii) $\quad \mathscr{P}_{\{\vee\}-BP} \quad = \quad \mathscr{NL}$,

(iii) $\quad \mathscr{P}_{\{\wedge\}-BP} \quad = \quad co\text{-}\mathscr{NL}$, and

(iv) $\quad \mathcal{P}_{\{\vee,\wedge\}-BP} \quad = \quad \mathcal{P} \, . \quad \blacksquare$

Furthermore, Theorem 3.2.1 along with the coincidence of the classes $\mathcal{NL}$ and $co\text{-}\mathcal{NL}$ [Im87, Sz87] yield the coincidence of the classes $\mathcal{P}_{\{\vee\}-BP}$ and $\mathcal{P}_{\{\wedge\}-BP}$.

COROLLARY 10.

Each polynomial size disjunctive branching program can be simulated by a polynomial size conjunctive branching program and vice versa. I.e.

$$\mathcal{P}_{\{\vee\}-BP} \quad = \quad \mathcal{P}_{\{\wedge\}-BP} \quad (= \ \mathcal{NL}) \, . \quad \blacksquare$$

Thus, four of the five classes of languages definable by means of polynomial size Ω-branching programs coincide with well-known nonuniform Turing machine complexity classes. However, although the fifth class $\mathcal{P}_{\{\oplus\}-BP}$ seems to be as interesting as the other ones, it has not been identified in the context of logarithmic space-bounded Turing machines up to now. We only know from Theorem 3.2.1 that it coincides with the class $\oplus\mathcal{L}$,

$$\oplus\mathcal{L} \quad = \quad \mathcal{L}(\{\oplus\}) \, ,$$

of languages nonuniformly acceptable by logarithmic space-bounded parity Turing machines.

Altogether we have proved:

THEOREM 3.2.2.

There are at most four complexity classes of languages related to polynomial size Ω-branching programs, $\Omega \subseteq \mathbb{B}_2$. These are the classes

$$\mathcal{P}_{BP} , \quad \mathcal{P}_{\{\vee\}-BP} = \mathcal{P}_{\{\wedge\}-BP} , \quad \mathcal{P}_{\{\oplus\}-BP} \quad \text{and} \quad \mathcal{P}_{\{\vee,\wedge\}-BP} .$$

They are interrelated in the following manner:

$$
\begin{array}{ccccc}
 & & \mathcal{NL} & & \\
 & & \| & & \\
 & & \mathcal{P}_{\{\wedge\}-BP} & & \\
 & \subseteq & \| & \subseteq & \\
 & & \mathcal{P}_{\{\vee\}-BP} & & \\
\mathcal{L} = \mathcal{P}_{BP} & & & & \mathcal{P}_{\{\vee,\wedge\}-BP} = \mathcal{P} . \\
 & & \mathcal{P}_{\{\oplus\}-BP} & & \\
 & \subseteq & \| & \subseteq & \\
 & & \oplus\mathcal{L} & & \\
\end{array}
$$

It is strongly recommended that all inclusions of this diagram are proper.

3.3. Bounded Width Ω-Branching Programs of Polynomial Size

In the following section we study polynomial size Ω-branching programs of bounded width for the purpose of characterizing the increase of computational power provided by Ω-branching programs, $\Omega \subseteq \mathbb{B}_2$, in the case of bounded width.

An Ω-branching program, $\Omega \subseteq \mathbb{B}_2$, is said to be *synchronous* if for each node v of P all paths from the source to v are of the same length. The *width* w of P is the maximal number of nodes in a level of P . A sequence of Ω-branching programs $\{P_n\}$ is said to be of *bounded width* if there is a constant c such that all P_n are of width $\leq c$.

An Ω-branching program P of length l is said to be in *normal form* if

(i)   each level j , $0 \leq j < l$, of P consists of the same number of nodes,

(ii) the source is the left most node of level 0 ,

(iii) each sink belongs to last level $l+1$, and

(iv) all nodes of a level are labelled either by the same input variable x_i , $1 \leq i \leq n$, or by Boolean functions $\omega \in \Omega$.

A straightforward argument shows that every Ω-branching program can be converted into a normal form Ω-branching program which accepts the same set   at the cost of doubling the width and multiplying the length by the minimum of the width and the number n of input variables. Since we are only interested in complexity results to within a constant (resp. a polynomial) factor we can assume our Ω-branching programs of bounded width to be in normal form. The same is true whenever we are dealing with polynomial size Ω-branching programs.

In the case of bounded width Ω-branching programs, $\Omega \subseteq \mathbf{B}_2$, the classification scheme of Section 3.1   can be condensed into the following two propositions. Proposition 11   extends the result of Proposition 3   to bounded width Ω-branching programs. It proves alternating $\{\vee,\wedge\}$-branching programs of bounded width to be the most powerful type of bounded width Ω-branching programs, $\Omega \subseteq \mathbf{B}_2$. Then, in Proposition 13, we show that $\{\vee,\wedge\}$-branching programs of bounded width are no more powerful than (ordinary) branching programs of bounded width. I.e. supplying bounded width branching programs with devices for evaluating   2-argument Boolean functions   does not increase their computational power.

PROPOSITION 11.

Let $\Omega \subseteq \mathbb{B}_2$. *Each* Ω*-branching program of width* w *and length* l *may be simulated by a* $\{\vee,\wedge\}$*-branching program of width* $\leq k_w \cdot w$ *and length* $\leq k_l \cdot l$ *for some constants* k_w , k_l $\in \mathbb{N}$.

PROOF.

Before we are going to prove this proposition we refer to the definition of the width of a circuit [Ba86]. Obviously, we can represent a circuit C as a rectangular array of gates. By introducing dummy nodes we can achieve that the edges entering a gate are from either inputs or gates on the immediately pre-vious row. Now, the *width* of a circuit is the minimum over such array representations of C of the maximal number of gates on a row.

Let us follow the proof of Proposition 3. Conversion of the given Ω-branching program P of width w and length l into the $(\Omega \cup \{sel\})$-circuit C_P causes at most a doubling of the width. The depth of C_P equals the length of P . Further, for each $\omega \in (\Omega \cup \{sel\})$ let w_ω and l_ω denote the width and the depth of a $\{\wedge,\vee\}$-circuit realization of ω , respectively. Then, the width of the $\{\wedge,\vee\}$-circuit C_P' simulating C_P is at most w_m times larger than that of C_P ,

$$w_m = \max \{w_\omega \mid \omega \in (\Omega \cup \{sel\})\} ,$$

and the length of C_P' is at most l_m times larger than that of C_P with

$$l_m = \max \{l_\omega \mid \omega \in (\Omega \cup \{sel\})\} .$$

Since the final conversion of C_P' into a $\{\wedge,\vee\}$-branching program P' simulating C_P' , and, hence, P does neither affect the width nor the length we are done with

$$k_w = 2 \cdot w_m \qquad \text{and} \qquad k_l = l_m . \blacksquare$$

Apart from simulating width restricted Ω-branching programs by alternating $\{\vee,\wedge\}$-branching programs with the same width restriction we have given mutual simulations of Boolean circuits of width w and alternating $\{\vee,\wedge\}$-branching program of width w in the proof of Proposition 11. Hence, width-restricted $\{\vee,\wedge\}$-branching programs and width-restricted Boolean circuits are computationally equivalent. However, our width-restricted alternating branching programs seem to be a more natural model of computation than that of width restricted Boolean circuits considered by Hoover, Barrington et.al. [Ba86].

COROLLARY 12.

(i) *Each alternating $\{\vee,\wedge\}$-branching program of width w and length l may be simulated by a Boolean circuit of width $2 \cdot w$ and depth $2 \cdot l$.*

(ii) *Each Boolean circuit of width w and depth l may be simulated by an alternating $\{\vee,\wedge\}$-branching program of width w and length $l+1$. $\blacksquare$*

While it is strongly conjectured that the complexity class of languages accepted by, say, polynomial size (ordinary) branching programs is properly contained in the class of languages accepted by polynomial size alternating branching programs the corresponding bounded width Ω-branching program classes coincide.

PROPOSITION 13.

An alternating $\{\vee,\wedge\}$-branching program of width w and length l may be simulated by an (ordinary) branching program of width 2^w and length l.

PROOF.

Let P be an alternating $\{\vee,\wedge\}$-branching program of width w and length l. We may simulate P by an (ordinary) branching program P' of width 2^w and length l which is constructed as follows: From the definition of an alternating branching program it can be concluded that during each computation every node of P is associated with a Boolean constant. Let us assume that the w nodes of the levels of P are associated with w Boolean variables $b_1,\ldots,b_w$ carrying these Boolean constants. If we let the 2^w nodes of the levels of P' represent the 2^w possible settings of these w Boolean variables, P' will be capable of simulating P level by level. A level j, $0 \le j < l-1$, of P is completely described by the two functions

$$f_j, g_j : [w] \longrightarrow [w],$$

where $[w] := \{1,\ldots,w\}$, which give the end points in level $j+1$ of the two edges leaving each node of level j. If level j is labelled by an input variable x_i then $f_j(v)$ is the end point of the edge starting in the node $v \in [w]$ of level j which corresponds to $x_i = 0$ where $g_j(v)$ is the end point of that edge starting in v which corresponds to $x_i = 1$. The last level $l-1$ which consists of non-sink nodes can be described by the two functions

$$f_{l-1}, g_{l-1} : [w] \longrightarrow \{0,1\}$$

which indicate the sinks to which the nodes of level $l-1$ are connected.

Starting with level 0 of P we, inductively, label level j, $0 \leq j < l$ of P and define functions f'_j and g'_j of P in the following way: If the nodes of level $l-1-j$ of P are labelled by an input variable x_i, $1 \leq i \leq n$, then we label level j of P by x_i, too. The two functions f'_j and g'_j,

$$f'_j, \quad g'_j: \quad 2^{[w]} \quad \longrightarrow \quad 2^{[w]},$$

describing level j of P are defined for $(b_1,...,b_w) \in \{0,1\}^W$ by

$$f'_j(b_1...b_w) = (b_{f_{l-1-j}(1)}, ..., b_{f_{l-1-j}(w)}) , \text{ and}$$

$$g'_j(b_1...b_w) = (b_{g_{l-1-j}(1)}, ..., b_{g_{l-1-j}(w)}) .$$

If the nodes of level $l-1-j$ of P are labelled by the Boolean functions $\omega_1,..., \omega_w \in \{\vee,\wedge\}$ then we label level j of P' by any one of the input variables x_i, $1 \leq i \leq n$, and define f'_j and g'_j for $(b_1,...,b_w) \in \{0,1\}^W$ by

$$f'_j(b_1,...,b_w) = g'_j(b_1,...,b_w) =$$

$$= (\omega_1(b_{f_{l-1-j}(1)}, b_{g_{l-1-j}(1)}),...., \omega_w(b_{f_{l-1-j}(w)}, b_{g_{l-1-j}(w)}) .$$

If we finally take the union of all nodes $(b_1,...,b_w) \in \{0,1\}^W$ with $b_1 = 1$ of level l of P' as 1−sink (that are the nodes of P which correspond to an accepting setting of the w Boolean variables of level 0 of P) and the union of the remaining nodes of level l of P' as 0−sink then, by means of an inductive argument, it is not difficult to prove that P and P' accept the same set. ∎

In particular, if the Ω−branching programs under consideration are of bounded width then, immediately from Proposition 11 and 13, we obtain

COROLLARY 14.

For each bounded width Ω-branching program, $\Omega \subseteq B_2$, there is a computationally equivalent bounded width (ordinary) branching program of the same length. ■

In the case of logarithmically bounded width we get

COROLLARY 15.

Every polynomial size Ω-branching program of logarithmic width, $\Omega \subseteq B_2$, may be simulated by an (ordinary) branching program of polynomial size. ■

Corollary 15 along with Hoover's simulation of width w branching programs by Boolean circuits of width $log\ w$ (cited and improved in [Ba86]) yields the following corollary.

COROLLARY 16.

Polynomial size (ordinary) branching programs and alternating branching programs of polynomial size and logarithmic width are of the same computational power. I.e. the complexity class $\mathscr{P}_{log\ \{\vee,\wedge\}-BP}$ of languages acceptable by (sequences of) alternating branching programs of logarithmic width coincides with the class $\mathscr{P}_{BP}$

$$\mathscr{P}_{log\ \{\vee,\wedge\}-BP}\ =\ \mathscr{P}_{BP}\ =\ \mathscr{L} .$$

PROOF.

Due to Corollary 12 the two concepts of Boolean circuits of width $O(w)$ and of alternating $\{\vee,\wedge\}$-branching programs of width $O(w)$ coincide. Hence Corollary 15 and the simulation result of Hoover yield the coincidence of $\mathcal{P}_{log\ \{\vee,\wedge\}-BP}$ and $\mathcal{P}_{BP}$. Theorem 1.1 implies the coincidence with $\mathcal{L}$. ∎

We conclude this section by summarizing the results obtained for bounded width Ω-branching programs of polynomial size.

THEOREM 3.3.

For each bounded width Ω-branching program, $\Omega \subseteq \mathbb{B}_2$, there is a computationally equivalent (ordinary) bounded width branching program. I.e. the complexity classes $\mathcal{P}_{bw\ \Omega-BP}$ of languages accepted by (sequences of) polynomial size Ω-branching programs of bounded width for each $\Omega \subseteq \mathbb{B}_2$ coincide with the class $\mathcal{NC}^1$ of languages computable by (sequences of) fan-in 2 Boolean circuits of depth $O(log\ n)$,

$$\mathcal{P}_{bw\ \Omega-BP} = \mathcal{NC}^1 , \quad \Omega \subseteq \mathbb{B}_2 . \blacksquare$$

Since it is strongly conjectured that $\mathcal{NC}^1$ is proper contained in $\mathcal{L} = \mathcal{P}_{BP}$ and, consequently, in $\mathcal{NL} = \mathcal{P}_{\{\vee\}-BP} = \mathcal{P}_{\{\wedge\}-BP}$, $\oplus \mathcal{L} = \mathcal{P}_{\{\oplus\}-BP}$, and $\mathcal{P} = \mathcal{P}_{\{\vee,\wedge\}-BP}$ it seems to be sure that, for each Ω-branching program, restricting the width of polynomial size Ω-branching program results in a definite restriction of their computational power.

3.4. Ω-Branching Programs of Quasipolynomial Size

The following section is devoted to the study of quasipolynomial size Ω-branching programs, $\Omega \subseteq \mathbb{B}_2$.

First, in Paragraph 3.4.1 we relate quasipolynomial size Ω-branching programs, $\Omega \subseteq \mathbb{B}_2$, to polylogarithmic space-bounded nonuniform Ω-Turing machines (Corollary 17) and quasipolynomial size bounded width Ω-branching programs to Boolean (fan-in 2) circuits of polylogarithmic depth (Corollary 19). Then, in Paragraph 3.4.2 we show that, whenever $\Omega \cup \{0, 1, id_I, id_I\}$ is not complete and $\Omega \neq \{\vee, \wedge\}$, all the classes $\mathcal{C}_{\Omega-BP}$, $\Omega \subseteq \mathbb{B}_2$, coincide (Theorem 3.4.2). Moreover, these classes coincide with the class $\Omega_{bw\ \Omega-BP}$ of languages acceptable by sequences of bounded width quasipolynomial size Ω-branching programs, a result which is very unlikely to be true in the corresponding case of polynomial size Ω-branching programs. The proof of Theorem 3.4.2 (properly that of Proposition 22) generalizes a well-known theorem of Savitch [Sa70].

3.4.1. COMPLEXITY CLASSES RELATED TO Ω-BRANCHING PROGRAMS OF QUASIPOLYNOMIAL SIZE

A sequence $\{P_n\}$ of Ω-branching programs, $\Omega \subseteq \mathbb{B}_2$, is said to be of *quasipolynomial* size if the size of P_n is bounded by $2^{(\log n)^{O(1)}}$. $\mathcal{C}_{\Omega-BP}$ denotes the class of all languages acceptable by sequences of quasipolynomial size Ω-branching programs.

Fully analogous to Theorem 3.2.1 we can relate quasipolynomial size Ω-branching programs and polylogarithmic (i.e. $(log\ n)^{O(1)}$) space-bounded nonuniform Ω-Turing machines.

COROLLARY 17.

For each $\Omega \subseteq \mathbb{B}_2$, quasipolynomial size Ω-branching programs and polylogarithmic space-bounded nonuniform Ω-Turing machines are of the same computational power. I.e.

$$\mathcal{Q}_{\Omega-BP} = \Omega\text{-}SPACE((log\ n)^{O(1)})\ /\ 2^{O((logn)^{O(1)})}, \quad \Omega \subseteq \mathbb{B}_2 ,$$

where $\Omega\text{-}SPACE(s(n))$ denotes the set of languages recognizable by $s(n)$ space-bounded Ω-Turing machines. ∎

If we consider the classes $\mathcal{Q}_{bw\ \Omega-BP}$, $\Omega \subseteq \mathbb{B}_2$, of languages acceptable by sequences of quasipolynomial size Ω-branching programs of bounded width then Proposition 11 and Proposition 13 yield:

COROLLARY 18.

For each quasipolynomial size Ω-branching program of bounded width, $\Omega \subseteq \mathbb{B}_2$, there is a computationally equivalent quasipolynomial size (ordinary) branching program of bounded width. Hence,

$$\mathcal{Q}_{bw\ \Omega-BP} = \mathcal{Q}_{bw\ BP}, \quad \Omega \subseteq \mathbb{B}_2 .\ \blacksquare$$

The following relation between quasipolynomial size branching programs of bounded width and Boolean (fan-in 2) circuits of polylogarithmic depth turns out to be useful in the study of quasipolynomial size unrestricted width Ω-branching programs, too.

COROLLARY 19.

Quasipolynomial size branching programs of bounded width and

Boolean (fan-in 2) circuits of polylogarithmic depth are of the same computational power. I.e.

$$\mathcal{Q}_{bw\ BP} = DEPTH((log\ n)^{O(1)})\ .$$

PROOF.

The proof follows immediately from Theorem 1.2 . ∎

Obviously, $\mathcal{Q}_{bw\ BP}$ contains the most important parallel complexity class, namely the class $\mathcal{NC}$ of all languages acceptable by polynomial size Boolean (fan-in 2) circuits of polylogarithmic depth.

COROLLARY 20.

$$\mathcal{NC} \subseteq \mathcal{Q}_{bw\ BP} = \mathcal{Q}_{bw\ \Omega-BP}\ .\ ∎$$

Let us summarize the above-mentioned properties of quasipolynomial size Ω-branching programs.

THEOREM 3.4.1.

(i) *For each $\Omega \subseteq \mathbf{B}_2$, quasipolynomial size Ω-branching programs and polylogarithmic space-bounded nonuniform Ω-Turing machines are of the same computational power. I.e*

$$\mathcal{Q}_{\Omega-BP} = \Omega-SPACE((log\ n)^{O(1)})\ /\ 2^{O((logn)^{O(1)})}\ ,\ \Omega \subseteq \mathbf{B}_2\ ,$$

(ii) *Quasipolynomial size Ω-branching programs of bounded width are no more powerful than quasipolynomial size (ordinary) branching programs of bounded width. I.e.*

$$\mathcal{Q}_{bw\ \Omega-BP} = \mathcal{Q}_{bw\ BP}\ ,\ \Omega \subseteq \mathbf{B}_2\ .\ ∎$$

3.4.2. RELATIONSHIP BETWEEN THESE CLASSES

Since the sizes of computationally equivalent Ω-branching programs coincide, to within a constant factor, due to the classification result of Theorem 3.1 each quasipolynomial size Ω-branching program, $\Omega \subseteq \mathbf{B}_2$, is computationally equivalent either to an ordinary branching program, to a disjunctive, to a conjunctive, to a parity or to an alternating branching program of quasipolynomial size. This, together with Corollary 18, yields

COROLLARY 21.

There are at most six complexity classes of languages re-lated to quasipolynomial size, unbounded and bounded width Ω-branching programs, $\Omega \subseteq \mathbf{B}_2$, respectively. These are the classes

$$Q_{bw\ \Omega-BP} \ = \ Q_{bw\ BP} \, , \quad \Omega \subseteq \mathbf{B}_2 \, ,$$

$$Q_{BP} \, , \quad Q_{\{\vee\}-BP} \, , \quad Q_{\{\wedge\}-BP} \, , \quad Q_{\{\oplus\}-BP} \quad \text{and} \quad Q_{\{\vee,\wedge\}-BP} \, .$$

They are interrelated in the following manner:

$$
\begin{array}{ccccccc}
& & & & \subseteq\ Q_{\{\vee\}-BP}\ \subseteq & & \\
Q_{bw\ BP} = Q_{bw\ \Omega-BP} & \subseteq & Q_{BP} & \subseteq & Q_{\{\oplus\}-BP} & \subseteq & Q_{\{\vee,\wedge\}-BP} \, . \ \blacksquare \\
& & & & \subseteq\ Q_{\{\wedge\}-BP}\ \subseteq & &
\end{array}
$$

While it is strongly conjectured that the corresponding classes of polynomial size Ω-branching programs do not coincide (with the exception of $\mathscr{P}_{\{\vee\}-BP}$ and $\mathscr{P}_{\{\wedge\}-BP}$) at least five of these six quasipolynomial size Ω-branching program classes do coincide. The proof of this fact is a generalization of the well-known theorem of Savitch [Sa70] which states that $NL \subseteq DSPACE((log\ n)^2)$.

PROPOSITION 22.

For $\omega \in \{\vee, \wedge, \oplus\}$ *each quasipolynomial size* $\{\omega\}-branching$ *program may be simulated by a Boolean (fan-in 2) circuit of polylogarithmic depth. Hence,*

(i) $\quad \mathcal{Q}_{\{\vee\}-BP} \quad \subseteq \quad DEPTH((log\ n)^{O(1)})$,

(ii) $\quad \mathcal{Q}_{\{\wedge\}-BP} \quad \subseteq \quad DEPTH((log\ n)^{O(1)})$, *and*

(iii) $\quad \mathcal{Q}_{\{\oplus\}-BP} \quad \subseteq \quad DEPTH((log\ n)^{O(1)})$.

PROOF.

Due to Corollary 19 it suffices to simulate each $\{\omega\}$-branching program, $\omega \in \{\vee, \wedge, \oplus\}$, of size $s(n)$ by a Boolean (fan-in 2) circuit of depth $O((log\ s(n))^2)$.

Let P be an $\{\omega\}$-branching program, $\omega \in \{\vee, \wedge, \oplus\}$, of size s . According to a construction mentioned in Section 1.2 we can assume without loss of generality P to be synchronous. Let $l = O(s)$ be the length of P . Since P is based on an acyclic graph we can enumerate its nodes by $1, ..., s$ in such a way that

- the source is numbered by 1 ,
- the 1-sink is numbered by s , and
- all nodes of a lower level are given smaller numbers than those of a higher level. (Consequently, each edge will always lead from a node with a lower number to a node with a higher number.)

Let $[i,j;k]_w$ be the number of paths of length k in P which will go in the case of an input w from node i to node j . Obviously, $[1,s;l]_w$ can be computed by means of the following recursion

$$[i,j;2k]_W = \sum_{i < m < j} [i,m;k]_W \cdot [m,j;k]_W \qquad (*)$$

in $\log l$ steps from the $[i,j;1]_W$, $1 \le i,j \le s$.

(i) Let P be a disjunctive branching program, i.e. $\omega = \vee$.

Clearly, P accepts an input w iff $sgn \; [1,s;l]_W = 1$ with

$$sgn \; [1,s;l]_W = \begin{cases} 1 & \text{if} \quad \# [1,s;l]_W \ge 1 ; \\ 0 & \text{otherwise}. \end{cases}$$

Applying sgn to $(*)$ we obtain

$$sgn \; [i,j;2k]_W = sgn \left(\sum_{i < m < j} [i,m;k]_W \cdot [m,j;k]_W \right)$$

$$= \bigvee_{i < m < j} sgn \; [i,m;k]_W \wedge sgn \; [m,j;k]_W .$$

By means of this recurrence $sgn \; [1,s;l]_W$ can be computed by an unbounded fan-in circuit of depth $O(\log l) = O(\log s(n))$ in the obvious way. By converting this circuit into a fan-in 2 circuit we obtain the desired depth $O((\log s(n))^2)$ circuit which simulates P.

(ii) Let P be a conjunctive branching program, i.e. $\omega = \wedge$.

Let $\neg P$ denote the equal sized $\{\vee\}$-branching program which can be obtained from P by replacing all $\wedge$-nodes by $\vee$-nodes and by letting the 1-sink be the 0-sink and the 0-sink be the 1-sink. Obviously, $\neg P$ accepts the complement $\bar{A}$ of the set A which is accepted by P.

Due to (i) we can simulate $\neg P$ by a fan-in 2 circuit of depth $O((\log s(n))^2)$. This circuit can easily be converted into a depth $O((\log s(n))^2)$ circuit which computes $\overline{A}$ instead of A and, hence, simulates P .

(iii) Let P be a parity branching program, i.e. $\omega = \oplus$.

LEMMA 23.
A $\{\oplus\}$-branching program P accepts an input $w \in \{0,1\}^n$ if the number of accepting paths under w is odd.

PROOF of the lemma:
We prove the lemma by induction on the length l of P .
The assertion is obvious for $l = 1$.

Now let P be a length l parity branching program and let the assertion be proved for all parity branching programs of length $\leq l-1$. If the source of P is labeled by a variable then the number of accepting paths in P under an input $w \in \{0,1\}^n$ equals that in the parity branching subprogram P' of P rooting in the endpoint of the first edge in P to be followed under w . Since P' is of length $l-1$ we are done. On the other hand, if the source of P is a $\oplus$-node then the number of accepting paths in P under w equals the sum of the accepting paths in both $\{\oplus\}$-branching subprograms P_0 and P_1 of P rooting in the two descendants of v_0 . Since these subprograms are of length $\leq l-1$ we can apply the induction hypothesis. Now by some easy computation it can be proved that the assertion is also true for P . $\square$

Lemma 23 implies that P accepts an input $w \in \{0,1\}^n$ iff

$$[1, s; l]_w \bmod 2 = 1 .$$

Applying $\bmod\ 2$ to the recurrence (*) yields

$$[i,j;2k]_W \ mod \ 2 \ = \ \left(\sum_{i \ < \ m \ < \ j} [i,m;k]_W \cdot [m,j;k]_W \right) \ mod \ 2$$

$$= \ \bigoplus_{i \ < \ m \ < \ j} \ ([i,m;k]_W \cdot [m,j;k]_W) \ mod \ 2$$

$$= \ \bigoplus_{i \ < \ m \ < \ j} \ ([i,m;k]_W \ mod \ 2) \wedge ([m,j;k]_W \ mod \ 2) \ .$$

This recurrence and the identity

$$a \oplus b = (\overline{a} \wedge b) \vee (a \wedge \overline{b})$$

allow to compute $[1,s;l]_W \ mod \ 2$ by means of an unbounded fan-in circuit of depth $O(log \ l) = O(log \ s(n))$ in the obvious way. By converting this circuit into a fan-in 2 circuit we obtain the desired depth $O((log \ s(n))^2)$ circuit which simulates P . ∎

Altogether we have obtained

THEOREM 3.4.2.

There are at most two complexity classes related to quasi-polynomial size Ω-branching programs, $\Omega \subseteq \mathbb{B}_2$, of bounded and unbounded width. These are the classes:

$$\mathcal{Q}_{\{\wedge\} -BP}$$

$$\|$$

$$\mathcal{Q}_{\{\vee\} -BP}$$

$$\|$$

$$DEPTH((log \ n)^{O(1)}) = \mathcal{Q}_{bw \ BP} \ = \mathcal{Q}_{bw \ \Omega-BP} = \mathcal{Q}_{BP} \ \subseteq \mathcal{Q}_{\{\vee,\wedge\} -BP} \ . \ \blacksquare$$

$$\|$$

$$\mathcal{Q}_{\{\oplus\} -BP}$$

Due to Theorem 3.4.1 this theorem can be restated as

COROLLARY 24.

For $\Omega \subseteq \mathbb{B}_2$, $\Omega \cup \{0, 1, id_l, id_r\}$ *not complete and* $\Omega \neq \{\vee,\wedge\}$, *polylogarithmic space-bounded nonuniform* Ω-*Turing machines and polylogarithmic space-bounded nonuniform (ordinary) Turing machines are of the same computational power.* ∎

3.5. Read-once-only Ω-Branching Programs of Polynomial size

Up to now superpolynomial and exponential lower bounds could be proved only for read-once-only branching programs (or similarly for real time branching programs) [We84], [Ža84], [A&86], [KW86], [Kr87]. In order to separate larger classes we generalize the concept of read-once-only branching programs by introducing and investigating read-once-only Ω-branching programs, $\Omega \subseteq \mathbb{B}_2$. Indeed, this approach proved to be quite successful since it enables us to separate the complexity classes $\mathcal{P}_{\Omega-BP1}$, $\Omega \subseteq \mathbb{B}_2$, related to polynomial size read-once-only Ω-branching programs (Theorem 3.5.4).

This section is organized as follows. In Paragraph 3.5.1 we relate polynomial size read-once-only Ω-branching programs, $\Omega \subseteq \mathbb{B}_2$, and logarithmic space-bounded nonuniform eraser Ω-Turing machines (Theorem 3.5.1).

In Paragraph 5.3.2 we confirm the classification result of Paragraph 3.1.2 to hold for read-once-only Ω-branching programs, too.

In Paragraph 5.3.3 we prove an exponential lower bound

(Lemma 33) and a polynomial upper bound (Lemma 35) for the problem of deciding whether a given Boolean matrix is a permutation matrix which both were obtained in cooperation with M. Krause and S.Waack. Moreover, in addition to the exponential lower bound of Lemma 4, Section 1.3 we give a polynomial upper bound for the problem of deciding whether a given undirected graph contains an odd number of triangles (Lemma 37).

Finally, in Paragraph 3.5.4 we use these lower and upper bounds for separating the complexity classes related to read-once-only Ω-branching programs (Theorem 3.5.4).

3.5.1. COMPLEXITY CLASSES RELATED TO POLYNOMIAL SIZE READ-ONCE-ONLY Ω-BRANCHING PROGRAMS

An Ω-branching program, $\Omega \subseteq \mathbf{B}_2$, is said to be *read-once-only* if every variable x_i , $1 \le i \le n$, is tested at most once on every computation path. By $\mathscr{P}_{\Omega-BP1}$ we denote the class of all languages $A \subseteq \{0,1\}^*$ whose restrictions $A^n = A \cap \{0,1\}^n$ will be accepted by read-once-only Ω-branching programs of polynomial size in n .

Following an approach of Ajtai et.al. [A&86] we have introduced in Section 1.3 the eraser Turing machine model and proved that the class of languages accepted by polynomial size read-once-only branching programs coincides with the class of all languages accepted by (nonuniform) logarithmic space-bounded (deterministic) eraser Turing machines . Now, investigating read-once-only Ω-branching programs, we generalize this approach by introducing eraser Ω-Turing machines, $\Omega \subseteq \mathbf{B}_2$, and relating polynomial size read-once-only Ω-branching programs and logarithmic space-bounded nonuniform eraser Ω-Turing machines.

An Ω-Turing machine M is called an *eraser Ω-Turing machine* if each path in the computation tree $T(M, w)$ of M on $w \in \{0,1\}^n+$ from the root to a leaf contains for all i, $1 \le i \le n$, at most one configuration properly depending on the i-th input bit. Indeed, eraser Ω-Turing machines generalize the concept of (deterministic) eraser Turing machines which erase an input bit after having read it.

An eraser Ω-Turing machine M is called $s(n)$ *space-bounded* if for all $w \in \{0,1\}^*$ the configurations included in the nodes of $T(M, w)$ consume at most space $s(|w|)$. Of special interest in our further considerations are the classes $L_e(\Omega)$ consisting of all languages acceptable by logarithmic space-bounded eraser Ω-Turing machines.

It is not difficult to see that these eraser Ω-Turing machines are generalizations of the common concepts of nondeterminism, co-nondeterminism and alternation for eraser Turing machines. If L_e, NL_e, co-NL_e and AL_e denote the classes of languages acceptable by logarithmic space bounded deterministic, nondeterministic, co-nondeterministic and alternating eraser Turing machines, then we obtain

PROPOSITION 25.

(i) $L_e(\emptyset)$ $=$ L_e ,

(ii) $L_e(\{\vee\})$ $=$ NL_e ,

(iii) $L_e(\{\wedge\})$ $=$ co-NL_e , and

(iv) $L_e(\{\wedge, \vee\})$ $=$ AL_e .

PROOF.

The proof is a consequence of the following two facts:

- eraser $\emptyset$-Turing machines, eraser $\{\vee\}$-Turing machines, eraser

{∧}-Turing machines, eraser {∨,∧}-Turing machines are deterministic, nondeterministic, co-nondeterministic and alternating Turing machines, respectively, with terminating computation paths, and

– every logarithmic space-bounded deterministic, nondeterministic, co-nondeterministic or alternating eraser Turing machine can be simulated by a logarithmic space-bounded deterministic, nondeterministic, co-deterministic or alternating eraser Turing machine, respectively, with terminating computation paths.

While the first fact is obvious the second can be proved in a way similar to Proposition 6 of this chapter. ∎

In order to relate read-once-only Ω-branching program classes, $\Omega \subseteq \mathbb{B}_2$, and eraser Ω-Turing machine classes we have to consider the *nonuniform counterparts* $\mathcal{L}_e(\Omega)$ of the eraser Turing machine classes $L_e(\Omega)$,

$$\mathcal{L}_e(\Omega) \;=\; L_e(\Omega) \,/\, n^{O(1)} \,,$$

consisting of languages $A \subseteq \{0,1\}^*$ for which there exists a polynomial length-restricted advice $\alpha : \mathbb{N} \longrightarrow \{0,1\}^*$ and a logarithmic space-bounded eraser Ω-Turing machine M such that M accepts $w\#\alpha(|w|)$ iff $w \in A$.

THEOREM 3.5.1.

Polynomial size read-once-only Ω-branching programs, $\Omega \subseteq \mathbb{B}_2$, and logarithmic space bounded nonuniform eraser Ω-Turing machines are of the same computational power. I.e.

$$\mathcal{P}_{\Omega-BP1} \;=\; \mathcal{L}_e(\Omega) \,, \quad \Omega \subseteq \mathbb{B}_2 \,.$$

PROOF.

The proof can be obtained from the proof of Theorem 3.2.1

with an additional remark concerning the read-once-only property of the considered Ω-branching programs and the eraser property of the corresponding Ω-Turing machines, respectively.

Since on every computation path a read-once-only Ω-bran-
ching program P tests each input bit x_i at most once the
simulating Ω-Turing machine will depend at most once on x_i
on each of its computation paths in the computation tree. Consequently, an eraser machine can do this job.

Reversely, since a given eraser Ω-Turing machine depends,
on each computation path, at most once from every input bit the
simulating Ω-branching program constructed in the proof of
Theorem 3.2.1 is read-once-only. ∎

Before we restate Theorem 3.5.1 in terms of deterministic,
nondeterministic, co-nondeterministic and alternating eraser
Turing machines let us briefly consider read-once-only alternating branching programs.

PROPOSITION 26.

Each alternating {∨,∧}*-branching program may be simu- lated
by an alternating read-once-only* {∨,∧}*-branching program of the
same size, to within a constant factor. Particularly we have*

$$\mathcal{P}_{\{∨,∧\}-BP1} = \mathcal{P}_{\{∨,∧\}-BP} = \mathcal{P} \ .$$

PROOF.

Using the simulations given in the proof of Proposition 3
we may first simulate a given {∨,∧}-branching program P by
the Boolean circuit C_P' , and second we may simulate this circuit C_P' by the alternating branching program P' . Since P'
can easily be seen to be read-once-only, and since both simulations keep the size unchanged, to within constant factors, we
are done. ∎

If $\mathcal{L}_e$, $\mathcal{NL}_e$, $co\text{-}\mathcal{NL}_e$, $\oplus \mathcal{L}_e$ and $\mathcal{AL}_e$ denote the non-uniform counterparts of the eraser machine classes L_e , NL_e , $co\text{-}NL_e$, $\oplus L_e$, and AL_e , then Theorem 3.5.1, together with Propositions 25 and 26 , yield

COROLLARY 27.

(i) $\quad \mathcal{P}_{BP1} \qquad = \qquad \mathcal{L}_e$,

(ii) $\quad \mathcal{P}_{\{\vee\}-BP1} \qquad = \qquad \mathcal{NL}_e$,

(iii) $\quad \mathcal{P}_{\{\wedge\}-BP1} \qquad = \quad co\text{-}\mathcal{NL}_e$,

(iv) $\quad \mathcal{P}_{\{\oplus\}-BP1} \qquad = \qquad \oplus \mathcal{L}_e$, and

(v) $\quad \mathcal{P}_{\{\vee,\wedge\}-BP1} \qquad = \qquad \mathcal{AL}_e \;=\; \mathcal{P}.\blacksquare$

3.5.2. CLASSIFICATION OF THE READ-ONCE-ONLY COMPLEXITY CLASSES

Most work for classifying read-once-only Ω-branching programs has already been done in Paragraph 3.1.2 . Since it can easily be seen that all the simulations given in the proofs of Propositions 2 – 5 respect the read-once-only property we obtain the following corollaries.

COROLLARY 28.

Let Ω_0 , $\Omega \subseteq \mathbf{B}_2$ with $\Omega_0 = \{0,1,id_l,id_r\}$. For each read-once-only $(\Omega \cup \Omega_0)$-branching program there is a computationally equivalent read-once-only Ω-branching program. $\blacksquare$

COROLLARY 29.

Let $\Omega \subseteq \mathbf{B}_2$ be a complete basis . For each read-onceonly

Ω-branching program there is a computationally equivalent alternating read-once-only $\{\vee,\wedge\}$-branching program. ∎

COROLLARY 30.

Let $\Omega \subseteq L = \{\neg_l, \neg_r, \leftrightarrow, \oplus\}$. Each read-once-only Ω-branching program P may be simulated by a parity read-once-only $\{\oplus\}$-branching program P of the same size, to within a constant factor. ∎

COROLLARY 31.

For each read-once-only $\{\neg\}$-branching program there is a computationally equivalent (ordinary) read-once-only branching program.∎

Altogether we obtain the following classification theorem:

THEOREM 3.5.2.

For each read-once-only Ω-branching program, $\Omega \subseteq B_2$, there is a computationally equivalent read-once-only Ω'-branching program with

$$\Omega' = \emptyset \ , \ \Omega' = \{\vee\} \ , \ \Omega' = \{\wedge\} \ , \ \Omega' = \{\oplus\} \ , \text{ or } \ \Omega' = \{\vee,\wedge\} \ . \ ∎$$

In particular, each polynomial size read-once-only Ω-branching program can be computationally equivalently represented either by an ordinary, by a disjunctive, by a conjunctive, by a parity or by an alternating read-once-only branching program of polynomial size. Due to the correspondence of Theorem 3.5.1 between polynomial size read-once-only Ω-branching programs and logarithmic space-bounded nonuniform eraser Turing machines we obtain

COROLLARY 32.

There are at most five complexity classes related to logarithmic space-bounded nonuniform eraser Ω-*Turing machines,* $\Omega \subseteq$ $\mathbf{B}_2$ *. These are the classes*

$$\mathcal{L}_e \, , \quad \mathcal{NL}_e \, , \quad co\text{-}\mathcal{NL}_e \, , \quad \oplus \mathcal{L}_e \, , \quad \mathcal{AL}_e = \mathcal{P} \, .$$

They are interrelated in the following manner:

$$
\begin{array}{ccccc}
 & \subseteq & \mathcal{NL}_e & \subseteq & \\
\mathcal{L}_e & \subseteq & \oplus \mathcal{L}_e & \subseteq & \mathcal{AL}_e = \mathcal{P} \, . \quad \blacksquare \\
 & \subseteq & co\text{-}\mathcal{NL}_e & \subseteq &
\end{array}
$$

3.5.3. SOME LOWER AND UPPER BOUNDS

In the following section we prove an $exp(\Omega(n))$ lower bound for the size of read-once-only disjunctive {∨}-branching programs which decide whether a given Boolean $n \times n$ matrix is a permutation matrix (Lemma 33). Additionally, we give a read-once-only conjunctive {∧}-branching program of size $O(n^2)$ which performs this task (Lemma 35). Both these results were obtained in cooperation with M.Krause and S.Waack [KMW88]. The present for of the proof of Lemma 33 was inspired by an argument of I.Wegener. Finally we give a polynomial size {⊕}-branching program which computes $\oplus cl_{n,3}$ from Section 1.3 (Lemma 37).

Let $F = \{f_n\}$ be the sequence of Boolean functions f_n defined on the set of Boolean $n \times n$ matrices A with

$f_n(A) = 1$ iff A is a permutation matrix, i.e. A contains exactly one 1 in every row and in every column.

Recall, that there is a 1-1 correspondence between $n \times n$ permutation matrices and n-permutations $\sigma \in S_n$

$$\sigma \longmapsto A^\sigma = (a_{ij}) \quad \text{with} \quad a_{ij} = \begin{cases} 1 & \text{if } j = \sigma(i) \\ 0 & \text{otherwise .} \end{cases}$$

LEMMA 33.

Every read-once-only disjunctive $\{\vee\}$-branching program which computes f_n is of size $2^{\Omega(n)}$.

PROOF.

At first we observe that in a disjunctive read-once-only branching program every computation path p_1 which starts at the source v_0 and leads to a non-sink node v can be combined with every computation path starting in v . I.e. if p_2 and p_3 are computation paths starting in v then $p_1 p_2$ as well as $p_1 p_3$ are computation paths starting in v_0 .

Now we let P be a disjunctive read-once-only branching program over $\{x_{11}, \ldots, x_{nn}\}$ which computes f_n . W.l.o.g. let n be even. We show

$$Size(P) \geq 2^n / 2\sqrt{n} .$$

Let p be an accepting path in P . Since P is read-once-only and since f_n is critical (i.e. for all $w \in f_n^{-1}(1)$ and all w' of Hamming distance 1 from w it holds $f_n(w') = 0$) there is exactly one permutation matrix $A = A^\sigma$ realized by p . For each permutation $\sigma \in S_n$ we fix an accepting path p_σ which belongs to the input A^σ .

To each $\sigma \in S_n$ we assign a non-sink node v_σ via p_σ in the following way: If $e = (v, v')$ is the $\frac{n}{2}$ - edge of p_σ with label 1, then

$$v_\sigma := v' \,.$$

We denote that part of p_σ which ends in v_σ by $p_\sigma^{(1)}$ and the remaining part of p_σ by $p_\sigma^{(2)}$. Let $V = \{v_\sigma \mid \sigma \in S_n\}$ and let $v \in V$ be an arbitrary node of V. Then we investigate S_V, the set of all permutations σ with $v_\sigma = v$,

$$S_V = \{\sigma \in S_n \mid v_\sigma = v\} \,.$$

By renumbering we can assume w.l.o.g. that $id \in S_V$ and x_{ii} is tested on $p_{id}^{(1)}$ iff $i \leq n/2$. Let $\sigma \in S_V$. According to our first observation the paths $p_{id}^{(1)} p_\sigma^{(2)}$ and $p_\sigma^{(1)} p_{id}^{(2)}$ are accepting paths in addition to p_{id} and p_σ. If x_{ij} is tested positively on $p_\sigma^{(1)}$ (or $p_\sigma^{(2)}$), and if $i \leq n/2$ and $j > n/2$ then $p_\sigma^{(1)} p_{id}^{(2)}$ (or $p_{id}^{(1)} p_\sigma^{(2)}$) is an accepting path where x_{ij} and x_{jj} (or x_{ii} and x_{ij}) are tested positively. However, this is a contradiction, since the disjunctive branching program P accepts only permutation matrices. Hence, S_V contains merely permutations σ with $\sigma(i) \leq n/2$ iff $i \leq n/2$. This implies

$$\#S_V \leq \left[\left(\begin{array}{c} n \\ 2 \end{array}\right)!\right]^2 \,.$$

Now we are done since

$$\#V \geq \frac{\#S_V}{\underset{v \in V}{max}\ \#S_V} \geq \frac{n!}{\left[\left(\frac{n}{2}\right)!\right]^2} = \left(\begin{array}{c} n \\ \frac{n}{2} \end{array}\right) \geq 2^n/2\sqrt{n} \,.$$

(due to Stirlings formula) and

$$Size(P) > \#V \,. \blacksquare$$

COROLLARY 34.

Every read-once-only conjunctive $\{\wedge\}$-branching program which computes $\neg f_n$ is of size $2^{\Omega(n)}$.

PROOF.

For every conjunctive read-once-only $\{\wedge\}$-branching program computing $\neg f_n$ we obtain a disjunctive read-once-only $\{\vee\}$-branching program of equal size computing $\neg(\neg f_n) = f_n$ if we replace the conjunctive $\wedge$-nodes by disjunctive $\vee$-nodes, the 1-sinks by 0-sinks and the 0-sinks by 1-sinks. Hence, Lemma 33 implies the corollary. ∎

LEMMA 35.

f_n can be computed by means of a read-once-only conjunctive $\{\wedge\}$-branching program of size $O(n^2)$.

PROOF.

Let e_n be the Boolean function defined by

$$e_n(w) = 1 \quad \text{iff} \quad |w| = 1$$

for $w \in \{0,1\}^n$. Using e_n we can write f_n as

$$f_n(x_{11},\dots,x_{1n},\dots x_{n1},\dots,x_{nn}) =$$

$$= \bigwedge_{i=1}^{n} e_n(x_{i1},\dots,x_{in}) \wedge \bigwedge_{j=1}^{n} e_n(x_{1j},\dots,x_{nj}) .$$

Since e_n can be computed by means of a read-once-only branching program of size $O(n)$ in a straightforward manner a conjunctive read-once-only $\{\wedge\}$-branching program of size $O(n^2)$ can be constructed which computes f_n. ∎

In analogy with Corollary 34 we obtain the following corollary to Lemma 35.

COROLLARY 36.

$\neg f_n$ *can be computed by means of a read-once-only disjunctive* {∨}*-branching program of size* $O(n^2)$. ∎

Finally, we give a polynomial upper bound for the size of a
read-once-only {⊕}-branching program which computes the
Boolean function ⊕ $cl_{n,3}$. This polynomial upper bound con-
trasts the exponential lower bound for the read-once-only branching program complexity of this function cited in Proposition
4 of Section 1.3. Recall from Section 1.3, ⊕ $cl_{n,3}$ = ⊕ $cl_{n,3}(x)$
computes 1 if the number of triangles in a given undirected
graph $G = G(x)$ is odd.

LEMMA 37.

⊕ $cl_{n,3}$ *can be computed by means of a read-once-only parity*
{⊕}*-branching program of size* $O(n^3)$.

PROOF.

Obviously, the following read-once-only branching program
$P_{u,v,w}$ decides whether the three nodes u, v, w $(1 \leq u < v <$
$w \leq n)$ of a given undirected graph $G = G((x_{ij}))$ $(1 \leq i < j \leq$
$n)$ constitute a triangle.

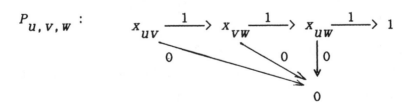

There are $\binom{n}{3}$ sets $\{u,v,w\} \subseteq \{1,...,n\}$ with $1 \leq u < v <$
$w \leq n$. Hence, if we replace the leaves of a binary tree of

$\oplus$-nodes of size $2 \cdot \binom{n}{3} - 1$ with all possible read-once-only branching programs $P_{u,v,w}$, u, v, $w \in \{1,...,n\}$ with $1 \le u < v < w \le n$, then we obtain a read-once-only $\{\oplus\}$-branching program of size $O(n^3)$ which computes $\oplus\ cl_{n,3}$. $\blacksquare$

3.5.4. SEPARATING COMPLEXITY CLASSES RELATED TO POLYNOMIAL SIZE READ-ONCE-ONLY Ω-BRANCHING PROGRAMS

Due to the lower and the upper bounds proved in Paragraph 5.5.3 for the problems $F = \{f_n\}$ and $\neg F = \{\neg f_n\}$ of deciding whether a given matrix is a permutation matrix and due to the lower and upper bounds for the problem $\{\oplus\ cl_{n,3}\}$ we can separate the five read-once-only Ω-branching program classes from each other in the following manner:

PROPOSITION 38.

(i) $\quad \mathscr{P}_{BP1} \overset{\subset}{\ne} \mathscr{P}_{\{\vee\}-BP1}$, $\mathscr{P}_{\{\wedge\}-BP1}$, $\mathscr{P}_{\{\oplus\}-BP1}$,

(ii) $\quad \mathscr{P}_{\{\vee\}-BP1} \not\subseteq \mathscr{P}_{\{\wedge\}-BP1} \not\subseteq \mathscr{P}_{\{\vee\}-BP1}$,

(iii) $\quad \mathscr{P}_{\{\wedge\}-BP1} \not\subseteq \mathscr{P}_{\{\oplus\}-BP1} \not\subseteq \mathscr{P}_{\{\wedge\}-BP1}$,

(iv) $\quad \mathscr{P}_{\{\oplus\}-BP1} \not\subseteq \mathscr{P}_{\{\vee\}-BP1} \not\subseteq \mathscr{P}_{\{\oplus\}-BP1}$, and

(v) $\quad \mathscr{P}_{\{\vee\}-BP1}$, $\mathscr{P}_{\{\wedge\}-BP1} \overset{\subset}{\ne} \mathscr{P}_{\{\vee,\wedge\}-BP1}$.

PROOF.
Trivially, we have

$$\mathscr{P}_{BP1} \subseteq \mathscr{P}_{\{\vee\}-BP1} , \mathscr{P}_{\{\wedge\}-BP1} , \mathscr{P}_{\{\oplus\}-BP1} \subseteq \mathscr{P}_{\{\vee,\wedge\}-BP1}$$

As in Section 3.5.3, we denote the problem of deciding whether a given Boolean matrix is a permutation matrix by $\{f_n\}$. $\{\oplus \; cl_{n,3}\}$ denotes the problem of deciding whether a given undirected graph contains an odd number of triangles.

(i) and (ii) can be easily obtained from the following element relations (1) to (8) proved in Paragraph 3.5.3 and cited in Section 1.3 :

(1) $\{f_n\}$ $\quad \notin \; \mathscr{P}_{\{\vee\}-BP1}$ (Lemma 33) and, consequently,

(2) $\{f_n\}$ $\quad \notin \; \mathscr{P}_{BP1}$.

(3) $\{f_n\}$ $\quad \in \; \mathscr{P}_{\{\wedge\}-BP1}$ (Lemma 35).

(4) $\{\neg f_n\}$ $\quad \notin \; \mathscr{P}_{\{\wedge\}-BP1}$ (Corollary 34) and, consequently,

(5) $\{\neg f_n\}$ $\quad \notin \; \mathscr{P}_{BP1}$.

(6) $\{\neg f_n\}$ $\quad \in \; \mathscr{P}_{\{\vee\}-BP1}$ (Corollary 36).

(7) $\{\oplus \; cl_{n,3}\} \notin \; \mathscr{P}_{BP1}$ (Lemma 4, Section 1.3), and

(8) $\{\oplus \; cl_{n,3}\} \in \; \mathscr{P}_{\{\oplus\}-BP1}$ (Lemma 37).

In order to obtain (iii) and (iv) remember that $\mathscr{P}_{\{\oplus\}-BP1}$ is closed under complementation. This is true since a read-once-only $\{\oplus\}$-branching program for a function h computes the negated function $\neg h$ after interchanging its 0-sink and its 1-sink. Since the relations (1) and (3) prove that the two classes $\mathscr{P}_{\{\vee\}-BP1}$ and $\mathscr{P}_{\{\wedge\}-BP1}$ are not closed under complementation they both cannot coincide with $\mathscr{P}_{\{\oplus\}-BP1}$.

Finally, (v) is a consequence of the fact that $\mathscr{P}_{\{\vee,\wedge\}-BP1}$ includes the non-coinciding sets $\mathscr{P}_{\{\vee\}-BP1}$, $\mathscr{P}_{\{\wedge\}-BP1}$. ∎

Since up to now only $\mathscr{P}_{BP1}$ has been separated by exponential lower bounds establishing these separations we have taken further steps in separating larger and larger complexity classes by means of exponential lower bounds.

Furthermore, the coincidence of $\mathcal{P}_{\{\vee\}-BP} = \mathcal{P}_{\{\wedge\}-BP}$ (Corollary 10) implies the separation of the read-once-only branching program classes $\mathcal{P}_{\{\vee\}-BP1}$ and $\mathcal{P}_{\{\wedge\}-BP1}$ from the branching program classes $\mathcal{P}_{\{\vee\}-BP}$ and $\mathcal{P}_{\{\wedge\}-BP}$, respectively. Hence, the read-once-only property provides a definite restriction of the computational power not only in the case of ordinary branching programs but also in the both cases of disjunctive and conjunctive branching programs.

COROLLARY 39.

(i) $\quad \mathcal{P}_{\{\vee\}-BP1} \overset{c}{\neq} \mathcal{P}_{\{\vee\}-BP}$, and

(ii) $\quad \mathcal{P}_{\{\wedge\}-BP1} \overset{c}{\neq} \mathcal{P}_{\{\wedge\}-BP}$.

PROOF.

(i) and (ii) follow immediately from Proposition 38 and from $\mathcal{P}_{\{\vee\}-BP} = \mathcal{P}_{\{\wedge\}-BP}$ proved in Corollary 10 of this chapter. ∎

Altogether, we have obtained

THEOREM 3.5.4.

The five read-once-only Ω-branching program classes

$$\mathcal{P}_{BP1}, \quad \mathcal{P}_{\{\vee\}-BP1}, \quad \mathcal{P}_{\{\wedge\}-BP1}, \quad \mathcal{P}_{\{\oplus\}-BP1} \quad and \quad \mathcal{P}_{\{\vee,\wedge\}-BP1}$$

are separated in the following manner:

$$\mathcal{P}_{BP1} \quad \begin{array}{c} \subset \\ \neq \end{array} \quad \begin{array}{c} \mathcal{P}_{\{\wedge\}-BP1} \\ \\ \\ \\ \mathcal{P}_{\{\wedge\}-BP1} \end{array} \quad \begin{array}{c} \subset \\ \neq \end{array} \quad \mathcal{P}_{\{\vee,\wedge\}-BP1} \quad (= \mathcal{P}_{\{\vee,\wedge\}-BP}) \ .$$

and

$$\mathcal{P}_{BP1} \quad \begin{array}{c} \subset \\ \neq \end{array} \quad \begin{array}{c} \mathcal{P}_{\{\vee\}-BP1} \\ \| \\ \mathcal{P}_{\{\oplus\}-BP1} \\ \| \\ \mathcal{P}_{\{\wedge\}-BP1} \end{array} \quad \blacksquare$$

Due to the correspondence of polynomial size read-once-only Ω-branching programs and logarithmic space-bounded eraser Ω-Turing machines (Corollary 27) we can restate Theorem 3.5.4 in terms of logarithmic space-bounded eraser Ω-Turing machines.

COROLLARY 40.

The five complexity classes

$$\mathcal{L}_e \ , \quad \mathcal{NL}_e \ , \quad co\text{-}\mathcal{NL}_e \ , \quad \oplus \mathcal{L}_e \quad \text{and} \quad \mathcal{P}_e \ = \ \mathcal{P}$$

related to logarithmic space bounded eraser Ω-Turing machines, $\Omega \subseteq \mathbb{B}_2$, are separated in the following manner:

$$\mathcal{NL}_e$$

$$\overset{\subset}{\neq} \qquad \qquad \overset{\subset}{\neq}$$

$$\mathcal{L}_e \qquad \qquad \overset{}{\cancel{\Vdash}} \quad \overset{}{\cancel{\Vdash}} \qquad \qquad \mathcal{AL}_e \ (= \mathcal{P}) \ ,$$

$$\overset{\subset}{\neq} \qquad \qquad \overset{\subset}{\neq}$$

$$co\text{-}\mathcal{NL}_e$$

and

$$\mathcal{NL}_e$$

$$\Vvdash$$

$$\mathcal{P}_{BP1} \quad \overset{\subset}{\neq} \quad \oplus \mathcal{L}_e \qquad . \quad \blacksquare$$

$$\Vvdash$$

$$co\text{-}\mathcal{NL}$$

Corollary 28   restated in terms of eraser Ω–Turing machines proves that the eraser concept causes definite restrictions of the computational power not only in the case of deterministic eraser Turing machines but also in the cases of nondeterministic and co-nondeterministic eraser Turing machines. Whereas Proposition 26   yields that alternating eraser Turing machines are of the same computational power than alternating non-erasing Turing machines.

COROLLARY 41.

Logarithmic space-bounded nondeterministic and co-nondeterministic nonuniform eraser Turing machines differ in their computational power and are definite less powerful than the corresponding non-eraser Turing machines. I.e.

$$\mathcal{NL}_e \ \neq \ co\text{-}\mathcal{NL}_e \ \text{ and }$$

$$\mathcal{NL}_e \ \overset{\subset}{\neq} \ \mathcal{NL} \ , \quad co\text{-}\mathcal{NL}_e \ \overset{\subset}{\neq} \ co\text{-}\mathcal{NL} \ = \ \mathcal{NL} \ . \ \blacksquare$$

APPENDIX

P-PROJECTION COMPLETE
GRAPH ACCESSIBILITY PROBLEMS

In this appendix we use the Ω-branching program characterizations of the (nonuniform) complexity classes

$$\mathcal{NC}^1 \, , \quad \mathcal{L} \, , \quad \mathcal{NL} = co\text{-}\mathcal{NL} \, , \quad \mathcal{P} \quad \text{and} \quad \mathcal{NP}$$

given in Chapters 1 to 3 to prove the completeness of a number of extremely restricted modifications of the *GRAPH-ACCESSIBILITY-PROBLEM* (*GAP*). Similar results are contained in [Me86,1], [Me87,1] and mainly in [Me87,4].

Apart from giving new insights into the capabilities of computations within certain complexity bounds, varying the complexity of one and the same problem makes the "differences" of the corresponding complexity classes more evident. For doing this we have chosen the GRAPH-ACCESSIBILITY-PROBLEM since it has proved to be of great importance for example in the study of the relations between deterministic and nondeterministic logarithmic space [Sa70], [Bu85], [Kri86], [Me87,1]. In unifying the approach we consider modifications of the GRAPH-ACCESSIBILITY-PROBLEM for switching graphs. Recall, a *switching graph* is a directed graph some of whose nodes v are *switchable* (that is they are equipped with certain switches s_v which by choice connect v with one of its successor nodes). Switching graphs are called *fully switchable* if all their nodes of outdegree at least 2 are switchable. If none of these nodes is switchable the switching graph is called *non-switchable*. The GRAPH-ACCES-SIBILITY-PROBLEM for switching graphs consists in deciding whether *all* edges reachable from a distinguished node v_0 belong to paths which lead to another distinguished node v_1. Since we are interested in considerations of very restricted GRAPH-ACCESSIBILITY-PROBLEMS we assume all our switching graphs to be *monotone* (i.e. all edges of the underlying graph whose vertex set is assumed to be enumerated lead from nodes with lower numbers to nodes with higher numbers) and of *outdegree 2* (i.e. at most two edges start in every node of the underlying graph).

The reducibility which we choose to work with is the stringent notion of the p-projection reducibility $\leq$ of Skyum and Valiant [SV81]. Thus our results are more comprehensive than when formulated in more liberal nonuniform reducibility concepts [CSV84].

In this appendix we prove the following theorem which covers only a small selection of the great variety of similar GRAPH-ACCESSIBILITY-PROBLEMS.

THEOREM.

The GRAPH-ACCESSIBILITY-PROBLEM for (monotone)

(1) *(switching) graphs of outdegree 1 is $\leq$-complete in* $\mathcal{L}$;

(2) *fully switchable switching graphs (of outdegree 2) is $\leq$-complete in $\mathcal{NL}$;*

(3) *non-switchable switching graphs (of outdegree 2) is $\leq$-complete in $co\text{-}\mathcal{NL} = \mathcal{NL}$;*

(4) *switching graphs of outdegree 2 is $\leq$-complete in $\mathcal{P}$;*

(5) *each of the above families of switching graphs is $\leq$-complete for $\mathcal{NC}^1$ if its bandwidth is bounded by a constant;*

(6) *switching graphs of outdegree 2 all of whose nodes of outdegree 2 are equipped with switches which can be coupled is $\leq$-complete in $\mathcal{NP}$;*

There is a general pattern which we will use to prove (1) to (6): In order to show that each of the modified GRAPH−ACCESSI−BILITY−PROBLEMS is ≤−hard in the cited complexity class we use the branching program description of that class which was given in the previous chapters. ≤−completeness can then be obtained by giving polynomial size branching programs of the appropriate type which solve the GRAPH−ACCESSIBILITY−PROBLEM under consid−eration. In the following we perform this program. □

Let us only mention that the problem of deciding whether the number of paths between v_0 and v_1 is odd is ≤−complete in $\oplus \mathcal{L}$ [Me87,1]. This can be proved similarly to the above theorem using the characterization of $\oplus \mathcal{L}$ by means of parity branching programs

Definitions

A *problem* is an infinite sequence of Boolean functions $F = \{f_n\}$ such that f_n has n variables. Via the usual corre−spondence of binary languages $A \subseteq \{0,1\}^*$ and sequences of Boolean functions $F(A) = \{f_n\}$, namely

$$w \in A \quad \text{iff} \quad f_{|w|} \in F(A) \quad \text{and} \quad f_{|w|}(w) = 1 ,$$

complexity classes can be regarded as classes of problems, too.

A problem $G = \{g_n\}$ is *p−projection reducible* to a problem $F = \{f_n\}$

$$G \leq F$$

if there is a function $p(n)$ bounded above by a polynomial in n , and if for every g_n of G there is a mapping

$$\pi_n : \{y_1,...,y_{p(n)}\} \longrightarrow \{x_1,\bar{x}_1,...,x_n,\bar{x}_n,0,1\}$$

such that

$$g_n(x_1,...,x_n) \ = \ f_{p(n)}(\pi_n(y_1),...,\pi_n(y_{p(n)})) \ .$$

From the definition the strictly nonuniform character of the p-projection reducibility concept becomes obvious.

A problem $F = \{f_n\}$ is $\leq$-*hard* in a complexity class $\mathcal{K}$ if all problems $G = \{g_n\}$ of $\mathcal{K}$ are p-projections of F . If F itself belongs to $\mathcal{K}$ then F is called $\leq$-*complete* in $\mathcal{K}$.

A *switching graph* $H = (G,S)$ consists of a directed graph $G = (V,E)$ and a set of *switches* $S = \{ s_v \mid v \in V \}$ assigned to some of the nodes of G with outdegree > 1. Such a switch $s_v \in S$ by choice connects the node $v \in V$ with one of its successor nodes. We call a node v of H *switchable* if there is a switch s_v assigned to v . A switching graph is called *fully switchable* if all of its nodes of at least outdegree 2 are switchable. If none of its nodes is switchable the switching graph is called *non-switchable*. Obviously, switching graphs of outdegree 1 are fully switchable as well as non-switchable. Briefly we will speak of graphs in this case.

Graph-theoretic properties attributed to a switching graph $H = (G,S)$ indicate properties of the underlying directed graph $G = (V,E)$. Generally, we assume G to be *monotone* (i.e. there is an enumeration of the nodes of G such that each edge leads from a node with a lower number to a node with a higher number) and of outdegree 2 (i.e. at most two edges start in every node). In accordance with the vertex enumeration given in a monotone graph we can encode G by the right upper part of the uniquely determined *adjacency matrix* $(a_{ij})_{1 \leq i < j \leq \# H = \# V}$

$$a_{ij} \ = \ \begin{cases} 1 & (i,j) \ \in \ E \ ; \\ 0 & \text{otherwise} \ . \end{cases}$$

The diagonal elements a_{ii}, $1 \leq i \leq \#V$, can be used for storing the information of whether a node i is switchable ($a_{ii} = 1$) or not ($a_{ii} = 0$). Since switches $s_i \in S$ are assigned only to nodes i of outdegree 2 possessing a left and a right successor node, *switch settings* of the switching graph $H = (G,S)$ can easily be described by a function

$$\delta : S \longrightarrow \{0,1\} .$$

δ indicates that the switch $s_i \in S$ connects i with its left ($\delta(s_i) = 0$) or its right ($\delta(s_i) = 1$) successor node.

Sometimes we consider switching graphs $H = (G,S)$ with *coupled switches*. Two switches s_i and s_j ($s_i, s_j \in S$) are *coupled in parallel* if the set of admissible switch settings is restricted to switch settings satisfying

$$\delta(s_i) = \delta(s_j) .$$

s_i and s_j are *orthogonally coupled* if the set of admissible switch settings is restricted to switch settings with

$$\delta(s_i) = (\delta(s_j) + 1) \bmod 2 .$$

We use a *coupling matrix* $(c_{ij})_{1 \leq i,j \leq \#V}$ to store the coupling information

$$c_{ij} = \begin{cases} 1 & \text{if } i \leq j \text{ and if } s_i \text{ and } s_j \text{ are coupled in parallel,} \\ & \text{or,} \\ & \text{if } i > j \text{ and if } s_i \text{ and } s_j \text{ are orthogonally coupled;} \\ 0 & \text{otherwise.} \end{cases}$$

Now we can formally describe the various GRAPH-ACCESSIBILI-TY-PROBLEMS. To keep things easy we treat all the GRAPH-ACCES-SIBILITY-PROBLEMS under consideration as sequences of *partial*

Boolean functions which are defined on the adjacency matrices (and coupling matrices, resp.) of the *appropriate* switching graphs. However, it is clear that these functions can be fully defined without increasing their complexity. E.g. by means of the branching programs given in Figures A.1 and A.2 it can be tested whether a given matrix $A = (x_{ij})_{1 \le i,j \le n}$ corresponds to a monotone graph of outdegree 1 or to a monotone graph of outdegree 2 , respectively. Furthermore, to make the argumentation as transparent as possible we use the index n in our GRAPH–ACCESSIBILITY–PROBLEMS to indicate the number of nodes of the switching graphs under consideration instead of the number of entrances in their adjacency matrix.

(1) 1GAP :

$1GAP = \{1GAP_n\}$ is the GRAPH–ACCESSIBILITY–PROBLEM for (monotone) graphs of outdegree 1 . It is defined by

$$1GAP_n : \{0,1\}^{n(n-1)/2} \overset{\supset}{\longrightarrow} \{0,1\}$$

$$(a_{ij})_{i<j} \longmapsto \begin{cases} 1 & \text{all edges reachable from vertex 1 belong to a path leading to } n \text{ in the monotone graph of outdegree 1 described by } (a_{ij})_{i<j} ; \\ 0 & \text{otherwise .} \end{cases}$$

(2) fsGAP :

$fsGAP = \{fsGAP_n\}$ denotes the GRAPH–ACCESSIBILITY–PROBLEM for fully switchable switching graphs

$$fsGAP_n : \{0,1\}^{n(n+1)/2} \overset{\supset}{\longrightarrow} \{0,1\}$$

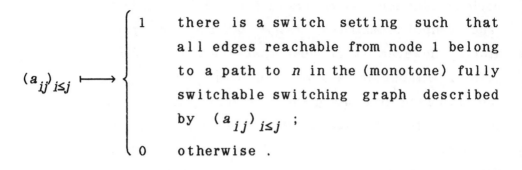

Obviously, *fsGAP* is equivalent to the problem of deciding whether there exist a path from vertex 1 to vertex n in an (ordinary) directed graph of outdegree 2 [Sa70], [CSV84], [Me86,1].

(3) nsGAP :

$nsGAP = \{nsGAP_n\}$ denotes the GRAPH−ACCESSIBILITY−PROBLEM for non−switchable switching graphs

$$nsGAP_n : \{0,1\}^{n(n+1)/2} \; \overset{\supset}{\longrightarrow} \; \{0,1\}$$

$$(a_{ij})_{i \leq j} \longmapsto \begin{cases} 1 & \text{all edges reachable from node 1 belong to a path to } n \text{ in the (monotone) non−switchable switching graph described by } (a_{ij})_{i \leq j} \; ; \\ 0 & \text{otherwise .} \end{cases}$$

Let us only mention that *nsGAP* is equivalent to *All−Path−GAP* which was considered in [Me87,1].

(4) sGAP :

$sGAP = \{sGAP_n\}$ denotes the GRAPH−ACCESSIBILITY−PROBLEM for switching graphs

$$sGAP_n \; : \quad \{0,1\}^{n(n+1)/2} \;\; \supseteq\!\!\longrightarrow \;\; \{0,1\}$$

$$(a_{ij})_{i \leq j} \longmapsto \begin{cases} 1 & \text{there is a switch setting such that} \\ & \text{all edges reachable from node 1 belong} \\ & \text{to a path to } n \text{ in the (monotone) swit-} \\ & \text{chable graph described by } (a_{ij})_{i \leq j}\; ; \\ \\ 0 & \text{otherwise .} \end{cases}$$

(5) bb-xGAP ($* \in \{1,\; fs,\; ns,\; s\}$) :

By $bb\text{-}{}^{*}GAP = \{bb\text{-}{}^{*}GAP_n\}$ we denote the GRAPH-ACCESSIBILITY-PROBLEM for (monotone) bounded bandwidth switching graphs of type $*$ ($* \in \{1,\; fs,\; ns,\; s\}$) . Recall, the *band-width* of a (monotone) graph G is the maximal difference $|i - j|$ of two adjacent nodes i and j of G.

(6) csGAP :

$csGAP = \{csGAP_n\}$ denotes the GRAPH-ACCESSIBILITY-PROBLEM for switching graphs H ,

$$H \;=\; (G,S) \;=\; \left((a_{ij})_{1 \leq i \leq j \leq n},\; (c_{ij})_{1 \leq i,j \leq n} \right) \;.$$

with coupled switches.

$$csGAP_n \; : \quad \{0,1\}^{n^2 + n(n+1)/2} \;\; \supseteq\!\!\longrightarrow \;\; \{0,1\}$$

$$H \longmapsto \begin{cases} 1 & \text{there is a switch setting satisfying the} \\ & \text{coupling conditions } (c_{ij}) \text{ such that each} \\ & \text{edge reachable from vertex 1 belongs to} \\ & \text{a path to } n \text{ in the switching graph des-} \\ & \text{cribed by } H = ((a_{ij})_{i \leq j}, (c_{ij})) ; \\ 0 & \text{otherwise .} \end{cases}$$

PROOF OF THE THEOREM

The proof of the Theorem follows from the following propositions.

PROPOSITION A.1.

The GRAPH-ACCESSIBILITY-PROBLEM 1GAP for (monotone) graphs of outdegree 1 is $\leq$-complete in $\mathcal{L}$.

PROOF.

In order to prove that *1GAP* is $\leq$-hard for $\mathcal{L}$ let $F = \{f_n\}$ be a problem of $\mathcal{L}$. Due to Corollary 1 of Section 1.1, for all n , there is a branching program P_n of polynomial size $p(n)$ which computes f_n . Using P_n we construct, for all inputs $(x_1, \dots, x_n) \in \{0,1\}^n$ of f_n , a (switching) graph $H(x_1, \dots x_n) = (\pi_n(x_{ij}))_{1 \leq i < j \leq p(n)}$ of outdegree 1 with

$$f_n(x_1, \dots x_n) = 1GAP_{p(n)}(H(x_1, \dots, x_n)) .$$

In detail, since P_n is based on an acyclic graph with $p(n)$ vertices we can enumerate its vertices by $1, 2, \dots, p(n)$ in such a way that

- the source is numbered by 1,

- the accepting 1-sink is numbered by $p(n)$, and
- each edge leads always from a node with a lower number to a node with a higher number.

Now, to every input $(x_1,...,x_n) \in \{0,1\}^n$ of f_n we assign a graph $H(x_1,...,x_n) = (\pi_n(x_{ij}))_{1 \leq i < j \leq p(n)}$ of size $p(n)$ with

$$
\pi_n(x_{ij}) = \begin{cases} x_k & \text{if vertex } i \text{ is labelled } x_k \text{ and if vertex} \\ & j \text{ is reached from } i \text{ if } x_k = 1 \text{ ;} \\ \overline{x}_k & \text{if vertex } i \text{ is labelled } x_k \text{ and if vertex} \\ & j \text{ is reached from } i \text{ if } x_k = 0 \text{ ;} \\ 0 & \text{otherwise .} \end{cases}
$$

By definition, $H(x_1,...,x_n)$ is of outdegree 1 . It is monotone due to the special nature of the enumeration of the vertices of P_n . Finally, since

$$f_n(x_1,...,x_n) = 1 \quad \text{iff} \quad P_n \text{ accepts } (x_1,...,x_n)$$

$$\text{iff} \quad 1GAP(H(x_1,...,x_n)) = 1$$

we obtain

$$F \leq 1GAP .$$

$1GAP$ is $\leq$-complete for $\mathcal{L}$, since $1GAP_n$, $n \in \mathbb{N}$, can be computed by the polynomial size branching program given in Figure A.3. □

In the following proposition we examine the GRAPH-ACCESSIBI-LITY-PROBLEM for fully switchable switching graphs. It is easy to see that this problem is equivalent to the problem of deciding whether there is a path in a directed graph of outdegree 2 which connects the nodes 1 and n [Me86,1]. However, due to the characterization of $\mathcal{NL}$ by means of 1-time-only-nondeterministic branching programs the proof of the classical completeness result [Sa70], [CSV84] becomes more obvious in our context and is strengthened for p-projection reductions.

PROPOSITION A.2.

The GRAPH−ACCESSIBILITY−PROBLEM fsGAP for (monotone) fully switchable switching graphs is $\leq$-complete in $\mathcal{NL}$.

PROOF.

For proving that fsGAP is $\leq$-hard for $\mathcal{NL}$ let $F = \{f_n\}$ be a problem of $\mathcal{NL}$. Due to Theorem 2.2.1 , for all n , there is a 1−time−only−nondeterministic branching program P_n of polynomial size $p(n)$ which computes f_n . Let the vertices of P_n be enumerated as in the proof of Proposition 1 . To every input $(x_1,...,x_n) \in \{0,1\}^n$ of f_n we assign a switching graph $H(x_1,...,x_n) = (\pi_n(x_{ij}))_{1 \leq i \leq j \leq p(n)}$ of size $p(n)$ with

$$
\pi_n(x_{ij}) = \begin{cases} 1 & \text{if vertex } i \text{ is labelled by a nondeter-} \\ & \text{ministic variable and if } (i,j) \text{ is an} \\ & \text{edge of } P_n \text{ ;} \\ x_k/\overline{x}_k & \text{if vertex } i \text{ is labelled } x_k \text{ and if ver-} \\ & \text{tex } j \text{ is reached from } i \text{ if } x_k \text{ is } 1/0 \text{ ;} \\ 0 & \text{otherwise .} \end{cases}
$$

for $i < j$, and

$$
\pi_n(x_{ii}) = \begin{cases} 1 & \text{if vertex } i \text{ is labelled by a} \\ & \text{nondeterministic variable;} \\ 0 & \text{otherwise ,} \end{cases}
$$

for $1 \leq i \leq p(n)$.

Obviously, $H(x_1,...,x_n)$ is a monotone switching graph of outdegree 2, all of whose nodes of outdegree 2 are switchable. Further, it holds

$$f_n(x_1,...,x_n) = 1 \quad \text{iff} \quad P_n \text{ accepts } (x_1,...,x_n) \, ,$$

i.e. iff there is a computation path q from source 1 to the accepting 1-sink $p(n)$ traced under $(x_1,...,x_n)$. Due to

$$q(i) \;=\; \begin{cases} 0 & i \text{ is connected in } q \text{ to its left successor node;} \\ 1 & \text{otherwise,} \end{cases}$$

we can define a switch setting δ in $H(x_1,...,x_n) = (\pi_n(x_{ij}))_{1 \leq i \leq j \leq p(n)}$

$$\delta : \; \{s_j \mid \pi(x_{ii}) = 1\} \longrightarrow \{0,1\}$$

by

$$\delta(s_j) \;=\; \begin{cases} 0 & \text{if } i \in supp \; q \text{ and } q(i) = 0 \, , \\ 1 & \text{otherwise,} \end{cases}$$

such that all edges reachable from node 1 belong to path q which goes to node $p(n)$.

Since each switch setting δ of a fully switchable switching graph $H(x_1,...,x_n)$ yields a monotone graph of outdegree 1 each edge reachable from node 1 belongs the same (uniquely determined) path. This path is an accepting path in P_n iff it goes from 1 to $p(n)$. Hence, there exists a switch setting δ which guaranties that each edge in $(H(x_1,...,x_n), \delta)$ which is reachable from node 1 belongs to a path to node $p(n)$ if and only if the 1-time-only-nondeterministic branching program P_n accepts $(x_1,...,x_n)$. That is

$$f_n(x_1,...,x_n) = 1 \quad \text{iff} \quad fsGAP_{p(n)}(H(x_1,...,x_n)) = 1$$

and, hence,

$$F \;\leq\; fsGAP \, .$$

Finally, *fsGAP* is ≤-complete for $\mathcal{NL}$, since it can be computed by the polynomial size 1-time-only-nondeterministic branching program given in Figure A.4 . In order to make things as simple as possible we merely give the stages S_i , $1 \le i < n$, from which the desired branching program is built and the connecting conditions of these stages. Since it is a property of all branching programs constructed in this section that from a stage S_i only stages S_j with $i < j$ can be reached, any S_i has to be taken only once in order to build the desired branching program. □

PROPOSITION A.3.

The GRAPH-ACCESSIBILITY-PROBLEM nsGAP for (monotone) non-switchable switching graphs is ≤-complete in co-$\mathcal{NL}$ = $\mathcal{NL}$.

PROOF.

The ≤-hardness of *nsGAP* will follow from Corollaries 9 and 11 of Paragraph 3.2.1 . There the complexity class co-$\mathcal{NL}$ = $\mathcal{NL}$ was characterized by means of polynomial size conjunctive branching programs. Let $F = \{f_n\}$ be a problem $\mathcal{NL}$. Taking a conjunctive branching program P_n of polynomial size $p(n)$ which computes f_n again we can assign to each input $(x_1,...,x_n) \in \{0,1\}^n$ of f_n a switching graph $H(x_1,...,x_n) = (\pi_n(x_{ij}))_{1 \le i \le j \le p(n)}$ of size $p(n)$ which is monotone, of outdegree 2, and without any switches:

$$
\pi_n(x_{ij}) = \begin{cases} 1 & \text{if vertex } i \text{ is an } \wedge\text{-node and if } (i,j) \\ & \text{is an edge of } P_n \text{ ;} \\ x_k/\overline{x}_k & \text{if vertex } i \text{ is labelled } x_k \text{ and if } j \\ & \text{is reached from } i \text{ if } x_k \text{ is } 1/0 \text{ ;} \\ 0 & \text{otherwise,} \end{cases}
$$

if $i < j$, and,

$$\pi_n(x_{ii}) = 0,$$

for all $1 \le i \le p(n)$.

Since P_n accepts an input $(x_1,...,x_n)$ iff all edges reachable from source 1 belong to paths to 1-sink $p(n)$ we obtain

$$f_n(x_1,...,x_n) = nsGAP_{p(n)}(H(x_1,...,x_n),$$

which proves that $nsGAP$ is $\le$-hard for $N\mathcal{L}$ ($= co-N\mathcal{L}$).

The $\le$-completeness of $nsGAP$ follows from the existence of a polynomial size conjunctive $\{\wedge\}$-branching program which computes $nsGAP_n$, $n \in \mathbb{N}$. Such a conjunctive branching program can be obtained from the 1-time-only-nondeterministic branching program of Figure A.4 by replacing each nondeterministic node by an $\wedge$-node. $\square$

PROPOSITION A.4.

The GRAPH-ACCESSIBILITY-PROBLEM sGAP for (monotone) switching graphs is $\le$-complete in $\mathcal{P}$.

PROOF.

From Theorem 3.2.1 and its Corollary 9 we know that any problem of $\mathcal{P}$ can be solved by means of polynomial size alternating branching programs. Let $F = \{f_n\}$ be a problem of $\mathcal{P}$ and let P_n be an alternating branching program of size $p(n)$ which is polynomial in n. Once more, for every input $(x_1,...,x_n) \in \{0,1\}^n$ of f_n we define a switching graph $H(x_1,...,x_n) = (\pi_n(x_{ij}))_{1 \le i \le j \le p(n)}$ such that

$$\pi_n(x_{ij}) = \begin{cases} 1 & \text{if vertex } i \text{ is labelled by } \vee \text{ or by } \wedge \\ & \text{and if } (i,j) \text{ is an edge of } P_n \text{ ;} \\ x_k/\overline{x}_k & \text{if vertex } i \text{ is labelled } x_k \text{ and if ver-} \\ & \text{tex } j \text{ is reached from } i \text{ if } x_k \text{ is } 1/0 \text{ ;} \\ 0 & \text{otherwise,} \end{cases}$$

for $i < j$, and

$$\pi_n(x_{ii}) = \begin{cases} 1 & \text{if the node } i \text{ is a } \{\vee\}\text{-node;} \\ 0 & \text{otherwise,} \end{cases}$$

for $1 \leq i \leq p(n)$.

Again one verifies

$$f_n(x_1,...,x_n) = sGAP_{p(n)}(H(x_1,...,x_n))$$

which implies that $sGAP$ is $\leq$-hard for $\mathcal{P}$.

Finally, $sGAP$ is $\leq$-complete in $\mathcal{P}$ since a polynomial size $\{\vee,\wedge\}$-branching program for $sGAP_n$, $n \in \mathbb{N}$, can be constructed from the 1-time-only-nondeterministic branching program of Figure A.4 by replacing all nondeterministic nodes by the branching program-component given in Figure A.5 . □

PROPOSITION A.5.

All the GRAPH-ACCESSIBILITY-PROBLEMS bb-1GAP , bb-fsGAP , bb-nsGAP and bb-sGAP for (monotone) graphs of outdegree 1 , for fully switchable switching graphs, for non-switchable switching graphs, and for switching graphs are ≤-complete in $\mathcal{NC}^1$.

PROOF.

Due to Theorem 3.3 each problem of $\mathcal{NC}^1$ can be solved by an Ω-branching program, $\Omega \subseteq \{\vee,\wedge\}$, of bounded width.

Starting with a bounded width (ordinary) branching program (resp. a disjunctive, a conjunctive, or an alternating branching program) we can proceed in a similar way as described in Proposition A.1 (resp. Proposition A.2, A.3, or A.4) . Since the resulting graph is of bounded bandwidth $bb-1GAP$ (resp. $bb-fsGAP$, $bb-nsGAP$, or $bb-sGAP$) is $\leq$-hard for $\mathcal{NC}^1$.

The $\leq$-completeness is a consequence of the existence of the respective type of polynomial size bounded width Ω-branching programs which compute $bb-1GAP_n$, $bb-fsGAP_n$, $bb-nsGAP_n$, or $bb-sGAP_n$, $n \in \mathbb{N}$. Since, treating switching graphs of bandwidth k , one has to test in each case merely k succeeding variables $x_{i,i+1}$,...., $x_{i,i+k}$ the desired bounded width Ω-branching programs can be obtained from those given in the proofs of Proposition A.1 to A.4. ∎

PROPOSITION A.6.

The GRAPH-ACCESSIBILITY-PROBLEM csGAP for (monotone) switching graphs with coupled switches is $\leq$-complete in $\mathcal{NP}$.

PROOF.

Due to Theorem 2.2.2 each problem $F = \{f_n\}$ of $\mathcal{NP}$ can be solved by (a sequences of) polynomial size nondeterministic branching programs. Once more, a nondeterministic branching program P_n of size $p(n)$ which computes f_n allows to construct a switching graph with coupled switches

$$H(x_1,...,x_n) = ((\pi_n(x_{ij}))_{1 \leq i \leq j \leq p(n)} , (\pi_n(c_{ij}))_{1 \leq i,j \leq p(n)})$$

for a given input $(x_1,...,x_n) \in \{0,1\}^n$ of f_n :

$$\pi_n(x_{ij}) = \begin{cases} 1 & \text{if vertex } i \text{ is labelled by a nondeter-} \\ & \text{ministic variable } z_k \text{ and if } (i,j) \text{ is} \\ & \text{an edge in } P_n ; \\ x_k/\overline{x}_k & \text{if vertex } i \text{ is labelled by } x_k \text{ and if } j \\ & \text{is reached from } i \text{ if } x_k \text{ is } 1/0 ; \\ 0 & \text{otherwise .} \end{cases}$$

For describing the coupling information correctly we have to make sure that, in the enumeration of the vertices of P_n , the left successor of a vertex i is numbered by a smaller number than the right successor . In order to do this we may have to negate some of the nondeterministic variables z_j assigned to vertices of P_n .

$$\pi_n(c_{ij}) = \begin{cases} 1 & \text{if } i \le j \text{ and if both nodes } i \text{ and } j \text{ are la-} \\ & \text{belled by the same literal of a nondeter-} \\ & \text{ministic variable } z_k , \text{ or} \\ & \text{if } i > j \text{ and if } i \text{ and } j \text{ are labelled by} \\ & \text{different literals of a nondeterministic} \\ & \text{variable } z_k ; \\ 0 & \text{otherwise .} \end{cases}$$

It is not difficult to check that $H(x_1,...,x_n)$ is a monotone switching graph of outdegree 2 with coupled switches and that

$$f_n(x_1,...,x_n) = csGAPH_{p(n)}(x_1,...,x_n)) .$$

Finally, $csGAP$ is $\le$-complete for $\mathcal{NP}$ since the nondeterministic branching program constructed in Figure A.6 is of polynomial size and computes $csGAP_n$, $n \in \mathbb{N}$. ∎

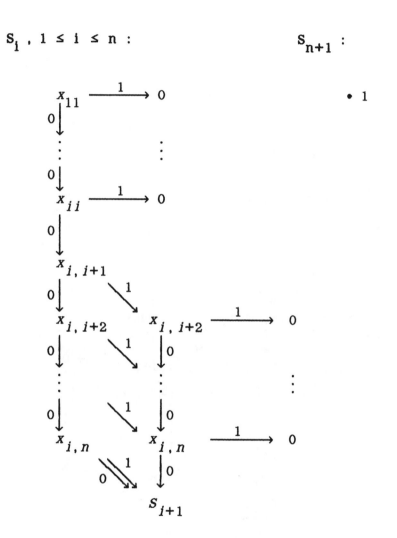

Figure A.1.

Stages from which an ordinary branching program can be buildt which tests whether a given graph $G = G(x_{ij})_{1 \leq i,j \leq n}$ is a monotone graph of outdegree 1 .

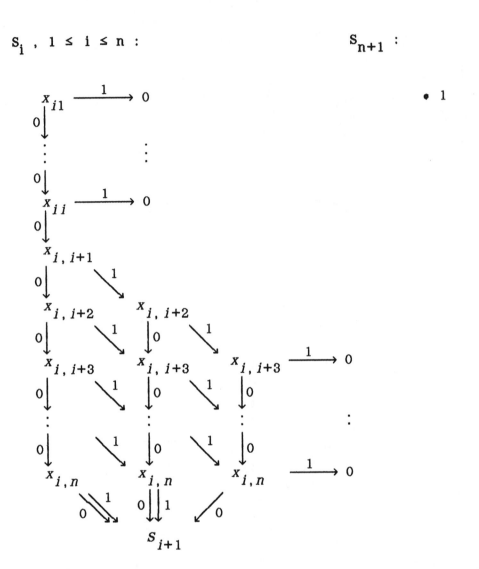

Figure A.2.

Stages from which an ordinary branching program can be buildt which tests whether a given graph $G = G(x_{ij})_{1 \le i,j \le n}$ is a monotone graph of outdegree 2 .

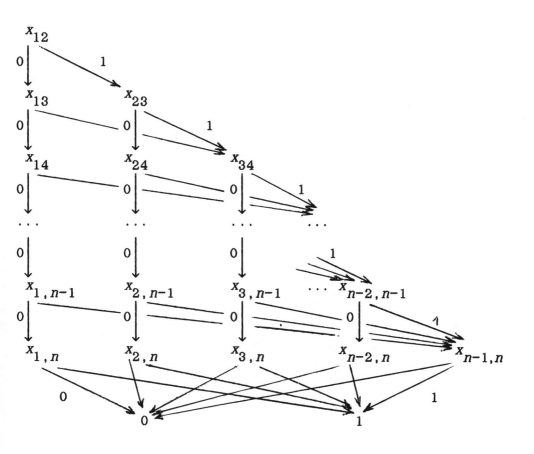

Figure A.3.

A Branching program which computes $1GAP_n$, $n \in \mathbb{N}$.

S_i , $1 \leq i < n$:

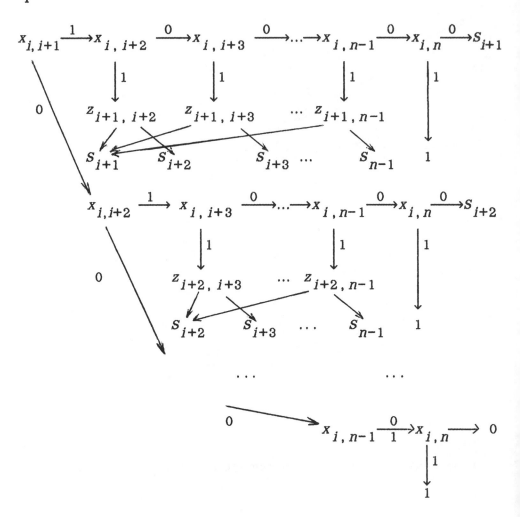

Figure A.4.

Stages of a 1-time-only-nondeterministic branching program which computes $fsGAP_n$, $n \in \mathbb{N}$. (Replacing all nondeterministic variables z_{kl} by $\vee$-nodes yields a disjunctive branching program for $fsGAP_n$.)

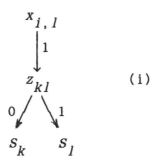

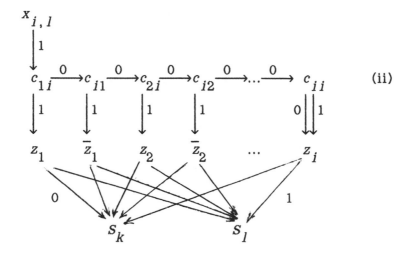

Figure A.6.

Replacing (i) by (ii) in the 1-time-only-nondeterministic branching program of Figure A.4 one obtains a nondeterministic branching program for $csGAP_n$, $n \in \mathbb{N}$. ∎

Figure A.5.

Replacing all nondeterministic nodes in S_j of Fig.A.4 by this branching program-component gives an alternating $\{\vee,\wedge\}$-branching program which computes $sGAP_n$, $n \in \mathbb{N}$.

References

[A&86] M.Ajtai,L.Babai,P.Hajnal,J.Komlos,P.Pudlak,V.Rödl, E.Szemeredi,G.Turan: Two lower bounds for branching programs, Proc. 18.ACM STOC (1986), 30–38.

[An85] A.E.Andreev: On a method of obtaining lower bounds for the complexity of individual monotone functions, Dokl. Akad. Nauk SSSR 282/5, 1033–1037.

[Ba86] D.A.Barrington: Bounded–width polynomial size branching programs recognize exactly those languages in NC^1, Proc. 18. ACM STOC, 1–5.

[BC80] A.Borodin, S.Cook: A time–space tradeoff for sorting on a general sequential model of computation, Proc. 12.ACM STOC (1980), 294–301.

[BDFP83] A.Borodin, D.Dolev, F.E.Fich, W.Paul: Bounds for width two branching programs, Proc. 15.ACM STOC (1983), 87–93.

[Bu82] L.Budach: personal communication.

[Bu85] L.Budach: Klassifizierungsprobleme und das Verhältnis von deterministischer und nichtdeterministischer Raumkomplexität. Sem.ber. Nr. 68, Sektion Mathematik, Humboldt–Univ. Berlin, 1985.

[CFL83] A.K.Chandra, M.L.Furst, R.J.Lipton: Multiparity protocols, Proc. 15.ACM STOC (1983), 94–99.

[CKS81] A.K.Chandra, D.C.Kozen,L.J.Stockmeyer: Alternation, J. of Ass. for Computing Machinery 28, 114–133.

[Co66] A.Cobham: The recognition problem for the set of perfect squares, Research paper RC–1704, IBM Watson Research Centre, 1966.

[CSV84] A.K.Chandra, L.Stockmeyer,U.Vishkin: Constant depth reducibility, SIAM J.Comput. 13, 2 (1984), 423–439.

[Du85] P.E.Dunne, Lower bounds on the complexity of 1–time only branching programs, Proc. FCT'85, LNCS 199, 90–99.

[FSS81] M.Furst, J.B.Saxe, M.Sipser: Parity, circuits, and the polynomial time hierarchy, Proc. 22. IEEE FOCS, 1981, 260–270.

[GP86] L.M.Goldschlager, I.Parberry: On the construction of parallel computers from various bases of Boolean functions, Theor. Computer Science 43 (1986), 43–58.

[Ha86]   J.Hastad: Improved lower bounds for small depth
           circuits, Proc. 18.ACM STOC (1986), 6–20.
[HU79]   J.E.Hopcroft, J.D.Ullman, Introduction to automata
           theory, languages, and computation, Addison-
           Wesley Publ. Comp. Inc., 1979.
[Im87]   N.Immerman: Nondeterministic space is closed under
           complement, Techn. Report 552, Yale Univ., 1987.
[KL80]   R.M.Karp, R.J.Lipton: Some connections between non-
           uniform and uniform complexity classes, Proc.
           12.ACM STOC (1980), 302–309.
[Kri86]   K.Kriegel, The space complexity of the accessi-
           bility problem for undirected graphs of log N
           bounded degree, Proc. MFCS'86, LNCS 233, 484–492.
[KW87]   K.Kriegel, S.Waack: Exponential lower bounds for
           real-time branching programs, Proc. FCT'87, LNCS
           278, 263–267.
[Kr86]   M.Krause: Exponential lower bounds on the complexi-
           ty of local and real-time branching programs, to
           appear in EIK 24 (1988) No. 3., 99–110.
[KMW88]   M.Krause, Ch.Meinel, S.Waack: Separating the eraser
           Turing machine classes $\mathcal{L}_e$, $\mathcal{NL}_e$, $co\text{-}\mathcal{NL}_e$ and $\mathcal{P}_e$,
           Proc. MFCS'88 (Karlovy Vary), LNCS 324, 405–413;
           to appear in Theor. Computer Science.
[La75]   R.E.Ladner: The circuit value problem is log space
           complete for P, SIGACT News 7, 18–20.
[Le59]   C.Y.Lee: Representation of switching functions by
           binary decision programs, Bell System Techn.
           Journal 38 (1959), 985–999.
[LF77]   R.E.Ladner, M.J.Fischer: Parallel prefix compu-
           tation, Journal ACM 27 (1980), 831–838.
[Ma76]   W.Masek: A fast algorithm for the string editing
           problem and decision graph complexity, M.Sc.
           thesis, MIT, 1976.
[Me86,1]   Ch.Meinel: p-projection reducibility and the com-
           plexity classes L(nonuniform) and NL(nonuniform),
           Proc. MFCS'86, LNCS 233, 527–535.
[Me86,2]   Ch.Meinel: Rudiments of a branching program based
           complexity theory, Preprint Nr. 127, Sekt.
           Mathematik, Humboldt-Univ. Berlin, 1986.
[Me87,1]   Ch.Meinel: Polynomial size Ω-branching programs and
           their computational power, to appear in Inf. and
           Computation.
[Me87,2]   Ch.Meinel: The nonuniform complexity classes $\mathcal{NP}$,
           $\mathcal{L}$ and $\mathcal{NL}$, EIK 23 (1987), No. 10/11, 545–558.
[Me87,3]   Ch.Meinel: The power of nondeterminism in polyno-
           mial-size bounded-width branching programs, Proc.
           FCT'87 (Kazan), LNCS 278, 302–309;
           Theor. Computer Science 62 (1988), 319–325.

[Me87,4]   Ch.Meinel: A complexity theoretic parade of net-
           work-flow-problems, ICOMIDC-Symp. on Math. and
           Computation, HoChiMinh city (1988);
           Fund. Informaticae XI (1988), 195-208.

[Me88]   Ch.Meinel: The power of polynomial size Ω-bran-
           ching programs, Proc. STACS'88 (Bordeaux), LNCS
           294, 81-90.

[MP75]   Muller,Preparata: Bounds to complexities of net-
           works for sorting and switching. Journal ACM 12
           (1975), 364-375.

[Ne66]   E.I.Nechipurok: On a Boolean function, Dokl. Akad.
           Nauk SSSR 169 No.4 (1966), 765-766.

[Pa78]   W.Paul: Komplexitätstheorie, Teubner Studienbücher
           Informatik, Stuttgart 1978.

[Po21]   E.Post: Introduction to a general theory of ele-
           mentary propositions, Am. J. Math. 43 (1921),
           163-185.

[Pu84]   P.Pudlak: A lower bound on complexity of branching
           programs, Proc. MFCS'84, LNCS 176, 480-489.

[PŽ83]   P.Pudlak, S.Žak: Space complexity of computations,
           Preprint Univ. of Prague, 1983.

[Ra85]   A.A.Razborov: A lower bound for the monotone net-
           work complexity of the logical permanent, Matem.
           Zametki 37/6.

[Ra86]   A.A.Razborov: Lower bounds on the size of boun-
           ded-depth networks over the basis $\{\wedge,\oplus\}$, Techn.
           Preprint Steklov Inst. Moskau, 1986.

[Sch85]   U.Schöning: Complexity and structure, LNCS 211
           (1986).

[Sa70]   W.Savitch: Relations between nondeterministic and
           deterministic tape complexities, J.Comp. and Sys.
           Sc. 4, 1970, 177-192.

[Sa76]   J.E.Savage: The complexity of computing, Wiley,
           1976.

[St76]   L.J.Stockmeyer: The polynomial time hierarchy,
           Theor. Computer Science 3, 23-33.

[SV81]   S.Skyum, L.G.Valiant: A complexity theory based on
           Boolean algebra, Proc. 22.IEEE FOCS (1981), 244-
           253.

[We84]   I.Wegener: On the complexity of branching programs
           and decision trees for clique functions, Techn.
           Report Univ. Frankfurt, 1984;
           JACM Vol. 35, No. 2 (1988), 461-471.

[We87]   I.Wegener: The Complexity of Boolean functions,
           Wiley-Teubner, Stuttgart, 1987.

[Va81]   L.Valiant: Reducibility by algebraic projections,
           L'Enseignement Mathm., t.XXVIII, fasc. 3-4, 253-
           268.

[Wr76]   C.Wrathall: Complete sets and the polynomial–time hierarchy, Theor. Computer Science 26, 287–300.

[Ya83]   A.C.Yao: Lower bounds by probabilistic arguments, Proc. 24.IEEE FOCS (1981), 244–253.

[Ya85]   A.C.Yao: Separating the poynomial–time hierarchy by oracles, Proc. 26.IEEE FOCS (1985), 1–10.

[Ža84]   S.Žak: An exponential lower bound for one–time–only branching programs, Proc. MFCS'84, LNCS 176, 562–566.

FCT   – Fundamentals of Computation Theory
FOCS   – Symp. on Foundations of Computer Science
LNCS   – Lecture Notes in Computer Science
MFCS   – Mathematical Foundations of Computer Science
STACS – Symp. on Theoretical Aspects of Computer Science
STOC   – Symp. on Theory of Computing

INDEX